161326678X

Two week loan

AF606259

Charges are made for late return.

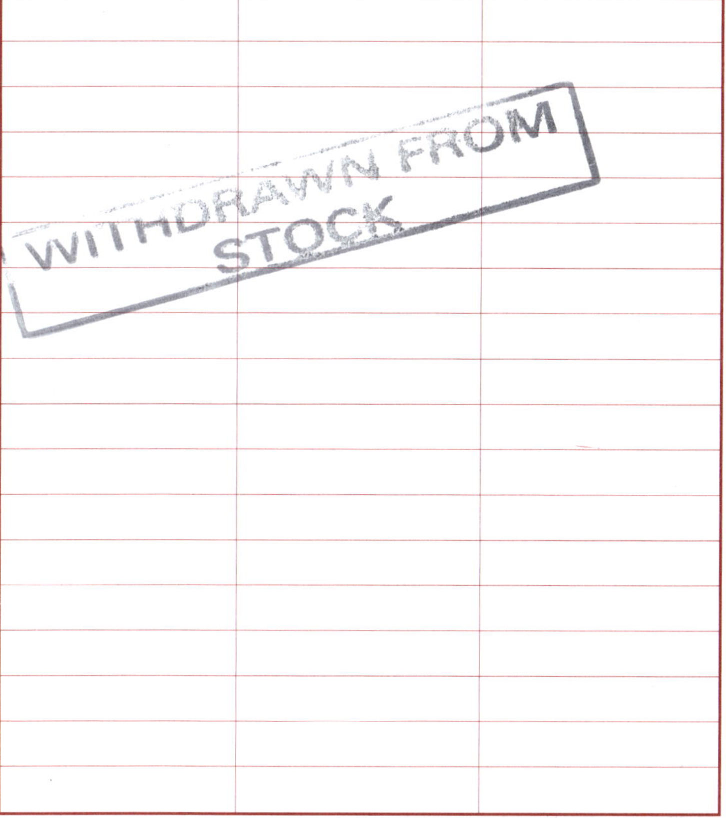

IS 239/0799

INFORMATION SERVICES PO BOX 430, CARDIFF CF10 3XT

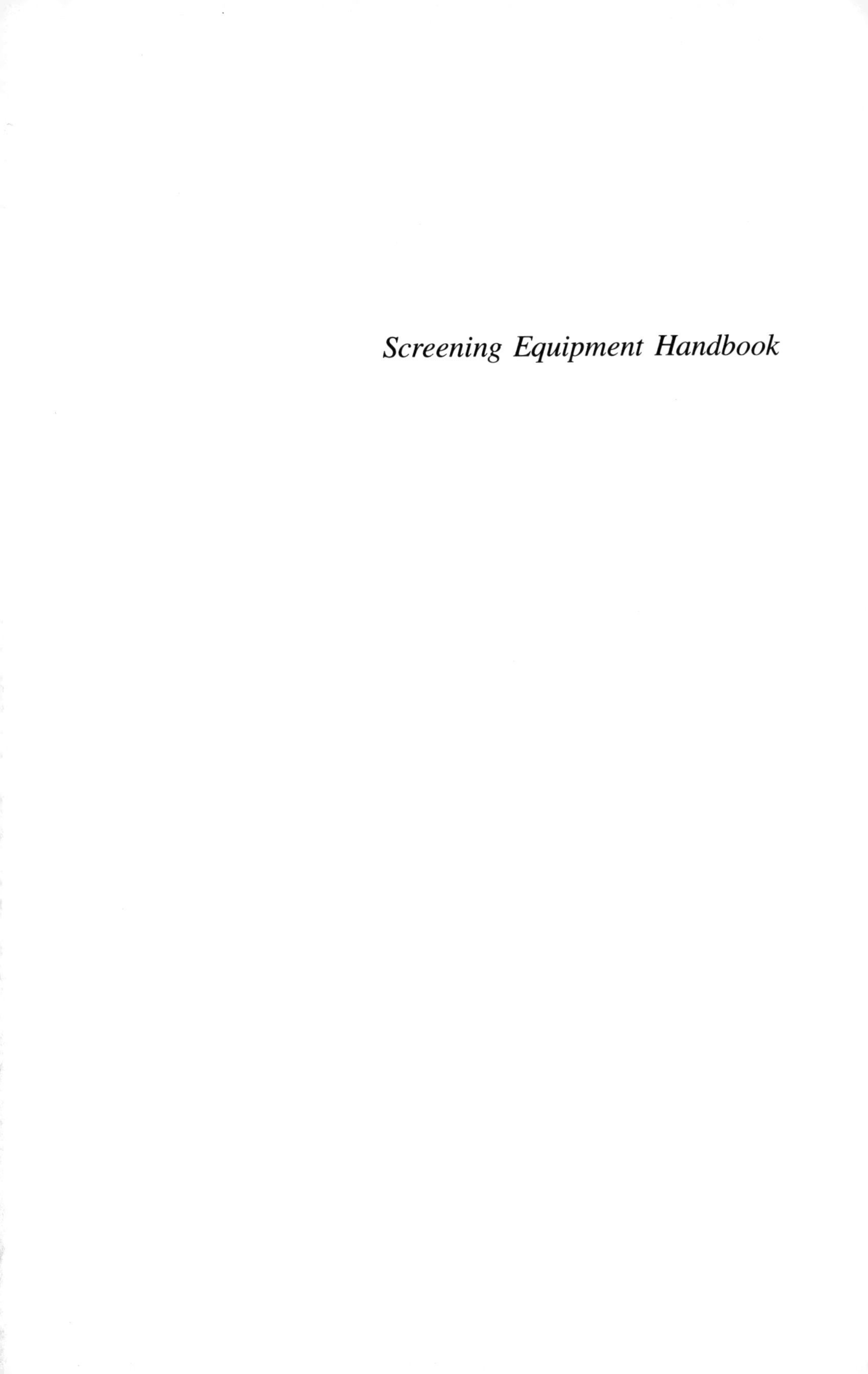

Screening Equipment Handbook

SCREENING EQUIPMENT HANDBOOK

for industrial and municipal water and wastewater treatment

Tom M. Pankratz

LANCASTER · BASEL

TD
439
·P2

Published in the Western Hemisphere by
Technomic Publishing Company, Inc.
851 New Holland Avenue
Box 3535
Lancaster, Pennsylvania 17604 U.S.A.

Distributed in the Rest of the World by
Technomic Publishing AG

© 1988 by Technomic Publishing Company, Inc.
All rights reserved

No part of this publication may be reproduced, stored in a retrieval system, or transmitted, in any form or by any means, electronic, mechanical, photocopying, recording, or otherwise, without the prior written permission of the publisher.

Published in cooperation with Lone Oak Publishing Company.

Printed in the United States of America
10 9 8 7 6 5 4 3 2 1

Main entry under title:
Screening Equipment Handbook: For Industrial and Municipal Water and Wastewater Treatment

A Technomic Publishing Company book
Bibliography: p. 253
Includes index p. 265

Library of Congress Card No. 88-50738
ISBN No. 87762-630-8

For MHP

PREFACE

The "simplicity" of most screens is deceiving, and their importance in overall plant performance is usually underestimated. During the preparation of this book I found no other reference that contained more than a few pages describing the various types of screening equipment currently available.

This book was written to consolidate the screening-related reference material that I have accumulated in the past 16 years. It is intended to be a comprehensive reference for the many professionals who design, manufacture, specify, purchase and operate screening equipment. I have tried to be as fair and objective as possible when reporting on the various types of equipment available.

Tougher operating requirements have resulted in many recent advances in water and wastewater screening equipment, yet some successful screen designs have gone almost unchanged since the early 1900's. The selection of the best screen for a specific application requires careful attention and usually involves the consideration of a number of engineering disciplines. Mechanical, civil, hydraulic, process, metallurgical and electrical engineering questions must be answered before the "right" decision can be made.

This book is divided into sections describing the major types of screening equipment. Sections 1, 2, and 3 deal with those screens generally used for raw water intakes. Section 4 covers screens primarily used in wastewater treatment applications, although trash rakes are also included in this section because of their similarity to bar screens. Section 5 covers fine screening, and Section 6 reviews microscreening in both water and wastewater applications.

Sections 7, 8, 9, and the appendixes contain a variety of information that should be of general interest. This information has been gathered from sources too numerous to mention, and represents the most interesting part of writing this book (the most difficult part was writing this preface).

I would like to thank my colleagues, Pat Conway, Gabriel Diamant, Bill Lauritch, John Lodholz, Robin Reddy, J. Ray Schrivener, Don Strow, Bill Unumb, and Jake Zelenietz who have each technically reviewed portions of this manuscript and made many valuable suggestions.

I am particularily greatful to the equipment manufacturers and the following individuals for providing the photographs and/or technical information that was necessary to complete this book: Richard J. Adams, L. Alleblas, Lincoln H. Berry, Peter Bigwood, Peter Blake, Buddy Brake, Roy Cheesman, Elaine Coffin, Steve Coombs, J.A. Coulson, Michael Drake, R.E. Fredrick, Larry Gargan, H.B. Haffer, Cindy Hendershot, Burd Hikes, Donald Klein, James Koski, Bill Lauritch, Larry Lehnert, R.A. Litkenhaus, Vernon D. Lucy, Gary D. Mackey, Gabriel Meunier, H.E. Meusching, J.K. Norg, Alfred F. A. Patzig, Dennis Pindar, Donald R. Portlock, Steven E. Ross, Stan Rudzinski, Dr. G. Schramm, James S. Schrier, Alan Smith, Rich Sommers, Mike Toepfer, Mark F. Tress, A. Verbandt, A. Wilcock, Wayne C. Wren.

I would also like to thank .....

Don Strow for taking the time to thoroughly answer my questions and for teaching me about Traveling Water Screens;
John Moloney for suggesting that I read the NYS Operators Manual;
Gil Koehler for showing me the importance of note-taking and scheduling;
Jack Matson for showing me the benefits of facilitative learning;
Doug Hagen and Alvin Van Black for teaching me not to concentrate on the obvious;
Gay Gibson for her willingness to follow sketchy instructions, poor penmanship and complicated text;
Marty Downey for introducing me to the Macintosh and insisting that I try MS Word;
and my mom and dad for setting a good example, encouraging me to be independent, and teaching me to never guess, to "look it up".

Finally, I would like to thank my wife Julie, and our children Chad, Sarah, Mikie, and Katie for their patience and support during the many hours, late nights, and long weekends that were required to prepare this manuscript.

Tom M. Pankratz
League City, Texas
20 March 1988

TABLE OF CONTENTS

SECTION 1 Traveling Water Screens 1

Applications 2

Thru Flow Traveling Screens 5

Dual Flow Traveling Screens 9

Thru Flow-to-Dual Flow Retrofit 17

Screen Variations 19

Screen Components 27

Design & Selection 54

SECTION 2 Other Intake Screens 66

Drum Screens 66

Revolving Disc Screens 71

Passive Screens 73

Radial Well Screens 76

SECTION 3 Fish Screens 78

Continuous Operation 79

Intake Location 81

Front Discharge Fish Screens 82

Rear Discharge Fish Screens 85

Flush Mounted Fish Screens 87

Platform Mounted Fish Screens 87

Radial Well Intake 88

Passive Screens 88

Horizontal Screens 88

Angled Fish Diversion Screen 90

Behavioral Barriers 90

SECTION 4 Bar Screens 92

Trash Rakes 93

Mechanically Cleaned Bar Screens 107

Other Bar Screening Equipment 149

Bar Screen Components 155

Bar Screen Sizing Procedures 159

SECTION 5 Fine Screens 167

Continuous Element Filter Screens 168

Rotary Screens 171

Disc-Type Fine Screens 178

Brush Raked Fine Screen 179

Static Screen 179

SECTION 6 Microscreening 182

Applications 183

Mesh 185

Drum & Support Frame 185

Backwash Spray System 187

Supplemental Mesh Cleaning 189

Drive & Controls 190

Sizing & Process Considerations 191

SECTION 7 Screen Control Systems 193

Hand-Off-Auto 193

Float Switch 194

Differential Level Controller 194

SECTION 8 Selection of Materials 198

Stainless Steel 199

Other Materials 200

Effects of Alloying Elements 203

SECTION 9 Corrosion Protection 206

Galvanic Corrosion 207

Protective Coatings 207

Cathodic Protection 208

APPENDIX 211

Abbreviations and Acronyms 212

Conversion Tables 220

Open Area Equations 226
Right or Left Hand? 227
Screening Equipment Manufactures 228

GLOSSARY 237
BIBLIOGRAPHY 253
INDEX 265

Knowledge is of two kinds;
we know a subject ourselves,
or we know where we can find
information upon it.

Samuel Johnson.

SECTION 1

TRAVELING WATER SCREENS

Traveling Water Screens, or Bandscreens, are automatically cleaned screening devices that are used to remove floating or suspended debris from a channel of water. Traveling Water Screens are usually used at raw, surface water intakes to protect pumping or other downstream equipment from objectionable debris.

Traveling Water Screens consist of a continuous series of wire mesh panels bolted to steel tray frames, or baskets, and attached to two matched strands of roller chain. The chain operates in a vertical path over head and footsprockets, carrying the trays down into the water, around the footsprockets, and back up through the water, over the headsprockets. As raw water passes through the revolving trays, debris is collected and retained on the upstream face of the wire mesh panels. The larger particles of debris are collected on a 2" to 3" wide lifting shelf that forms the lower, or trailing edge of the tray frame.

The debris-laden trays are lifted out of the flow and above the operating floor where a high pressure water spray is directed outward through the mesh to remove impinged debris. The spray wash water and debris is collected in a trough for further disposal.

The path of the chain and tray assemblies is guided within a track that also forms a skeleton-like frame to support the screen structure. A labyrinth seal between the screen frame structure and channel wall prevents debris from passing around the screen. Each tray is furnished with two end seal plates to seal the openings between the ends of the tray and the chain guide tracks.

Traveling Water Screens are usually not operated until a specified differential headloss or time lapse has occurred. Although some applications may be operated more frequently, even continuously, based on specific design or application requirements.

There are two distinct types of Traveling Water Screens available; the "Thru Flow" and the "Dual Flow". In the Thru Flow design, the screening surface is oriented perpendicular to the flow with only the ascending baskets utilized as available screening area. In the Dual Flow design, the screening surface is oriented parallel to the flow utilizing both the ascending and descending screens as active screening area.

Traveling Water Screens can be manufactured to suit virtually any raw water screening application. Screens are available with tray widths that range from 2 to 14 feet, wire mesh openings from 1/8" to 3/4", and channel depths up to 100 feet or more.

APPLICATIONS

Electric Power Generation

Nuclear and fossil fueled steam electric generating plants require approximately 500 to 1500 gallons of water per minute per megawatt of rated capacity. This water is primarily used for cooling purposes in the plants condensers, and must be screened prior to use to prevent clogging of tubes and openings in the surface and jet condensers as well as interfering with the proper operation of the circulating and condenser pumps. Traveling Water Screens are used to fulfill this screening requirement at virtually all power plants that use a surface cooling water source.

Many power plants discharge the condenser exhaust directly to a lake, ocean or river that serves as a heat sink. A large power plant that utilizes this "once through" approach in its cooling water system may require as many as forty eight individual Traveling Water Screens. Most newer power plants recycle their condenser exhaust through cooling towers. Approximately 3% to 5% of the recirculating water must be continually replaced to makeup for loses due to evaporation, drift, and blowdown. Such a "closed cycle" cooling water system greatly reduces plant intake water requirements, and consequently, the number of Traveling Water Screens.

Hydro-electric power stations may use Traveling Water Screens to screen the pumped recirculating water on a fish ladder system. This is a relatively low flow application compared to the other previously described power plant screen uses, and usually requires two or three screens.

Multiple Traveling Water Screen Installation
Courtesy of FMC Corporation

Wood Products Industry

Pulp and Paper Plants are intensive users of raw water, requiring up to 20,000 gallons of water per ton of product. Many pulp, paper and saw mills generate their own electricity and use Traveling Water Screens to screen raw condenser feed water and process water prior to its use.

Most mills process the logs that are used as raw materials for their product. These logs are conveyed from the woodyard to large debarking drums by means of a six to fifteen foot deep log flumes. Prior to being recycled, the water passes through a grit chamber followed by Traveling Water Screens to remove bark, leaves and twigs that have fallen off or been removed during handling.

Some paper plants and kraft mills utilize Traveling Water Screens in their waste treatment process, prior to clarification, to remove floating and suspended fibers or particles that could otherwise hinder sedimentation.

Miscellaneous Manufacturing

Steel Mills, Petro-Chemical Plants, Refineries, and other large manufacturing plants may also require the use of Traveling Water Screens. In these manufacturing plant applications, Traveling Water Screens are most often used to screen cooling water for purposes associated with steam generation.

Traveling Water Screens may also be used to screen fire water, process cooling water, boiler feed water, potable water and process water.

Irrigation Projects

Irrigation and other open-channel water distribution projects in the western and southwestern United States often use a Traveling Water Screen to protect pumping equipment at remote pump stations.

Water Treatment Plants

Large surface water treatment and desalination plants employ Traveling Water Screens to screen raw intake water as the first step in the treatment process. Inadequate screening may result in damage to downstream pumping or other treatment equipment.

THRU FLOW TRAVELING WATER SCREENS

Thru Flow Traveling Water Screens, also referred to as "direct", "single", or "uni" flow screens, are the most common type of raw water intake screen in the United States. Over 12,500 Thru Flow Traveling Water Screens have been manufactured since 1895.

The terminology used to describe this screen refers to the waters flow pattern through it. A Thru Flow Screen is installed in a channel with the screening surfaces oriented perpendicular to the water flow. Raw water passes through the ascending and descending screen trays, respectively, utilizing the ascending trays on the upstream(front) side of the screen as active screening surfaces. As debris is elevated out of the flow it is removed by a high pressure water spray before the cleaned trays begin their descent on the downstream (rear) side of the screen.

Debris is prevented from passing around the screen through the use of vertical guides that also serve to locate the screen in the well. Sealing plates are mounted on the both ends of each screen tray to prevent solids from passing around the sides of the trays as they rotate around the footsprocket and ascend in the upstream chain guide tracks.

Thru Flow screens typically utilize 80 to 90% of the channel width as effective screening area, and occupy less than 5'-6" of channel length (in direction of the flow). The straight-through flow pattern usually means that no elaborate forebay is required, and that any required pumps can be located relatively close to the screens.

Thru Flow Traveling Water Screen
Courtesy of FMC Corporation

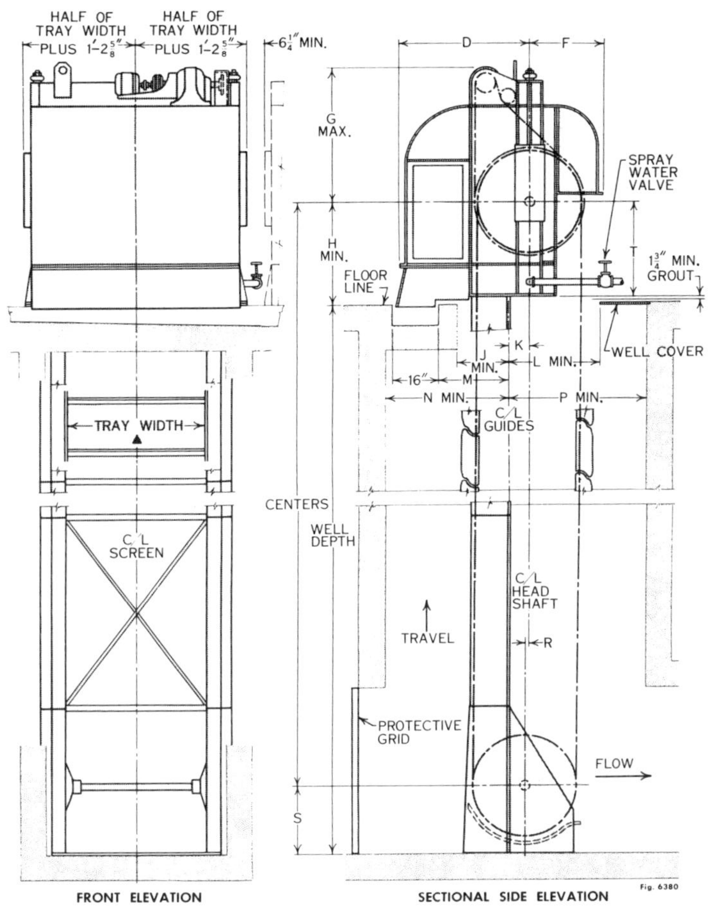

Model screen	D	F	G	H	J	K	L	M	N	P	R	S	T
	FEET AND INCHES												
45A	4-4⅛	2-9¾	4-11½	3-5½	1-8¾	0-11¾	3-7½	2-2¾	3-9	5-0	0-1½	2-5½	3-3
46A	3-8⅝	2-2⅜	4-1½	3-0	1-6¼	0-6⅞	2-7¼	2-0¼	3-6½	3-11¾	0-1¼	2-0	2-9

Thru Flow Traveling Water Screen

Courtesy of FMC Corporation

Traveling Water Screen Front/Back View
Courtesy of McGinnes/Royce Corporation

The primary advantage of the Thru Flow type screen is the reduced cost of intake construction and equipment. Intakes that incorporate Thru Flow screen designs tend to be compact, with simple and uncomplicated flow patterns.

The potential for debris "carry-over" is the most important disadvantage of the Thru Flow Traveling Water Screen. If the spray system does not effectively remove debris from the ascending trays, the debris will be carried over into the flow on the downstream side of the screen. This defeats the purpose of the screen and may result in damage to downstream equipment. The only protection from carry-over problems is proper design, maintenance and monitoring of the screen and its spray wash system.

Because only the half of the total submerged screening area (ascending trays) is utilized, a Thru Flow screen installation will require more/larger screens to pass a given flow than a Dual Flow installation. This generally results in higher operating and maintenance costs.

DUAL FLOW TRAVELING WATER SCREENS

Dual Flow Traveling Water Screens were developed in the 1920's and have been used extensively in Europe and Asia. They are being applied on an increasingly frequent basis in the United States. Dual Flow and Thru Flow Traveling Water Screen differ in the waters flow pattern through the screen. The screening surfaces of Dual Flow screens are oriented parallel to the water flow with raw water passing through both the ascending and descending screen trays.

This flow pattern offers some significant advantages over Thru Flow designs. One advantage is that the entire submerged screen surface is utilized as active screen area. This means that a Dual Flow screen of a given width will pass almost twice as much water, at the same velocity, as a Thru Flow screen of the same basket width.

Dual Flow Traveling Water Screen
Courtesy of Hubert Water Systems BV

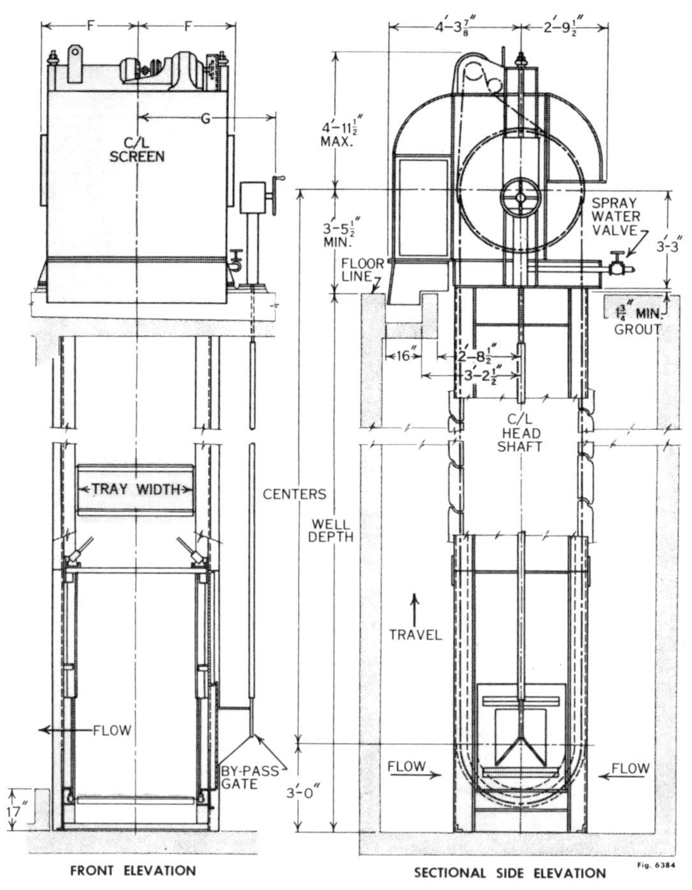

Tray width	F	G
FEET AND INCHES		
3-0	2- 8⅝	4- 1⅛
3-6	2-11⅝	4- 4⅛
4-0	3- 2⅝	4- 7⅛
4-6	3- 5⅝	4-10⅛
5-0	3- 8⅝	5- 1⅛
5-6	3-11⅝	5- 4⅛
6-0	4- 2⅝	5- 7⅛
6-6	4- 5⅝	5-10⅛
7-0	4- 8⅝	6- 1⅛
7-6	4-11⅝	6- 4⅛
8-0	5- 2⅝	6- 7⅛

Double-Entry Dual Flow Screen

Courtesy of FMC Corporation

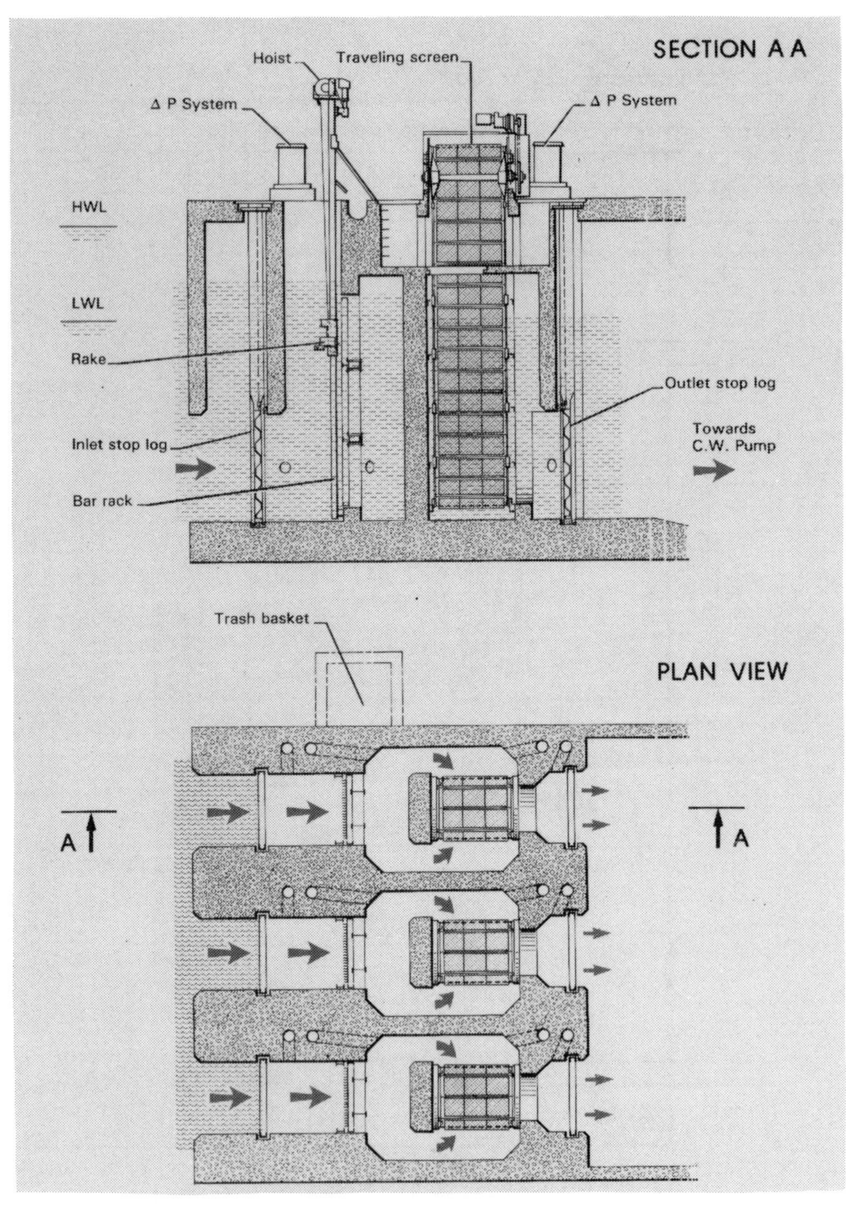

Double-Entry Dual Flow Intake
Courtesy of Beaudrey Corporation

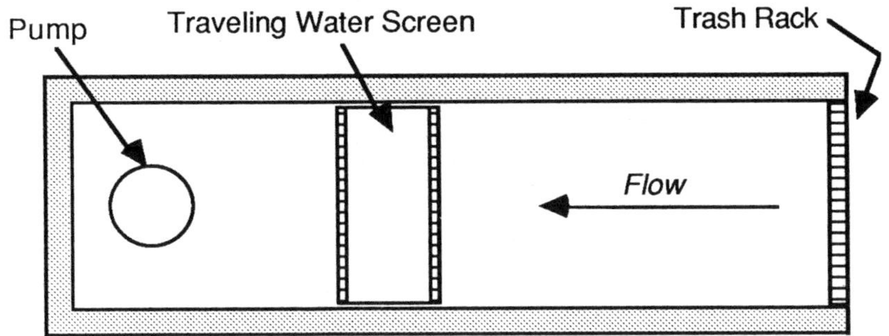

Thru Flow Screen

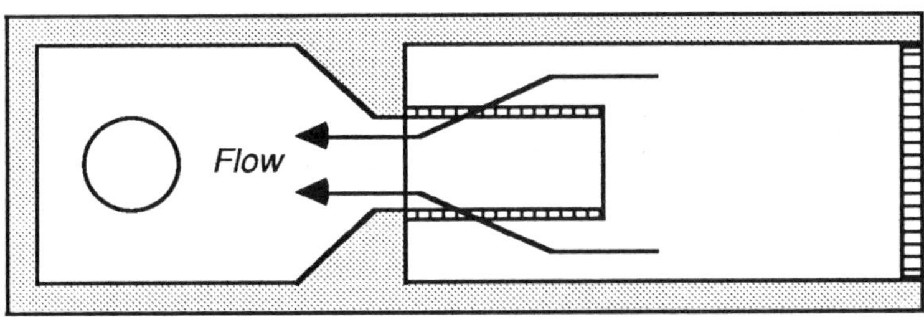

Double-Entry/Single-Exit Dual Flow Screen

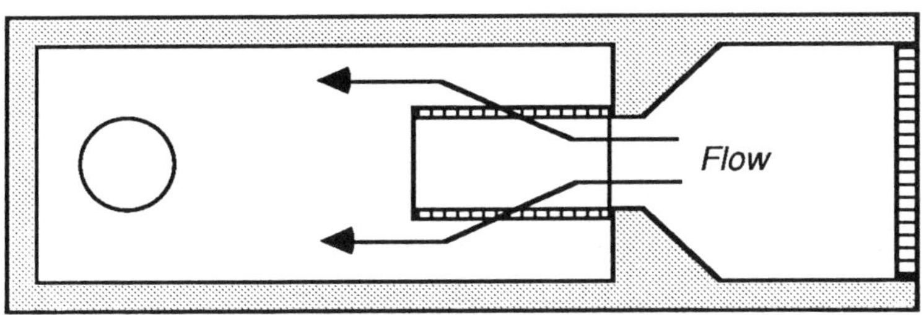

Single-Entry/Double-Exit Dual Flow Screen

DUAL FLOW SCREEN ARRANGEMENTS

Center Flow Traveling Water Screen
Courtesy of Passavant-Werke

Even more important is the fact that the Dual Flow designs virtually eliminate the possibility of debris carry-over. The debris remains on the upstream side of the screening surface on both the ascending and descending trays at all times. The boot section design of Dual Flow screens eliminates the boot sealing problems associated with Thru Flow designs.

The main disadvantage of the Dual Flow Traveling Water Screen is the increased construction costs due to the larger concrete channel and support structure. The screen orientation and the flow pattern of water will also require careful hydraulic consideration to prevent high localized velocities resulting from abrupt changes in the direction of the water flow.

Applications where severe debris loading conditions are anticipated should also consider that debris collected on the descending tray run will block the screen for the remainder of the operating cycle. This may require continuous operation and/or structural modifications to insure that the screen will withstand higher, and more frequent differential headloss conditions.

Although the industry standard tray width for Thru Flow screens is 10 feet, most Dual Flow screens have a tray width of 5 to 7 feet. The narrower tray widths result in an overall reduction in the weight of the tray/chain assembly, and consequently, a reduction in sprocket, chain, and bearing wear. The shorter unsupported span of the structural members in a narrower tray also makes them less vulnerable to damage from a headloss.

Modifications can be made to some Dual Flow designs that will allow the use of mesh openings as small as 1 mm, or less.

It is generally accepted that the initial cost of an intake utilizing a Thru Flow screen is less than that of a Dual Flow. Based on the type and quantity of debris, the higher first cost may be offset by the reduced operating costs generally associated with a Dual Flow design.

Double-Entry/Single-Exit

Double-Entry/Single-Exit Traveling Water Screens are the most common type of Dual Flow screen. Also called "double", "twin", and "central" flow screens, these units are mechanically similar to a Thru Flow screen that has been rotated 90 degrees in the channel.

The Double-Entry Dual Flow screen draws water through the face of both the ascending and descending trays. Screened water exits through an opening in the center of the screen structure.

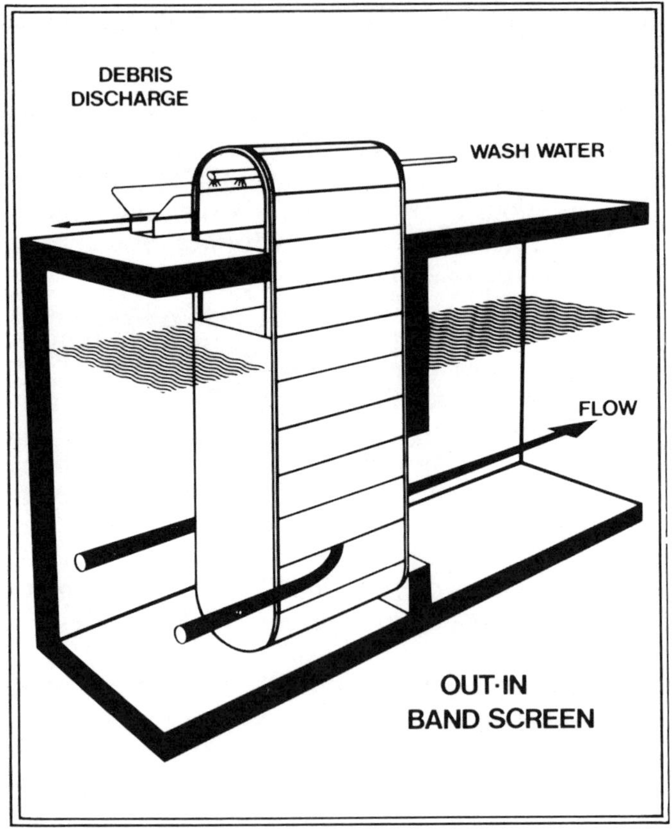

Courtesy of Hawker Siddeley Brackett

Single-Entry/Double-Exit

Water enters Single-Entry/Double-Exit screen through an opening in the center of the screen frame and exits outward through the trays. These screens are also referred to as a "center" or "internal" flow units, and are available for channel depths to 34 feet.

Most Single-Entry Dual Flow screens have semi-circular, or involute shaped screen trays. This tray shape provides approximately 60% more screening area per tray, and helps retain debris during its ascent. Additional retaining plates may be added to the trays to further enhance debris carrying ability.

The radius of the chain guide track in the foot terminal may be more than twice that of the headsprocket, further increasing the submerged screen area.

The Single-Entry Dual Flow screen has a debris spray located external to the screen, providing a downward debris discharge as the trays reach their uppermost elevation.

THRU FLOW-TO-DUAL FLOW RETROFIT

Most existing Thru Flow screen installations can be modified, or retrofit to use a Double-Entry Traveling Water Screen. The modification consists of the installation of a special wall plate mounted perpendicular to the flow, in place of the existing screen. The Dual Flow screen is then lowered into the well, with trays parallel to the flow, on the upstream side of the wall plate. An inlet opening in the wall plate allows screened water to pass to the pumps.

An alternative arrangement uses a specially constructed screen mainframe that includes a wall plate made as an integral part of the frame with extensions or "wings" that fit into the existing embedded guides.

A Thru Flow-to-Dual Flow retrofit provides increased flexibility and

Single Entry Dual Flow Screens
Courtesy of Davy Bamag GmbH

Single-Entry Dual Flow Screen, housing removed
Courtesy of Davy Bamag GmbH

may allow the operator to solve a specific operational problem. If an existing Thru Flow screen can be replaced with a Dual Flow unit of a similar tray width, the resulting increase in available screening area will provide a reduced velocity through the mesh. The user may also elect to install a screen with finer mesh, or a narrower tray width, without increasing the velocity.

Typical Power Plant Screen Installation
Courtesy of FMC Corporation

SCREEN VARIATIONS

Down Spray Screens. Several manufacturers offer this variation of Thru Flow and Double-entry Dual Flow Traveling Water Screens. This screen design utilizes a headshaft arrangement that results in the screening surface of the descending trays achieving a near horizontal orientation directly over the trough at the point of discharge.

Some designs utilize multiple headshafts to direct the trays to a horizontal position. The ascending trays pass over a set of drive sprockets before rounding a second set of idler sprockets to begin their descent. As they are guided around the idler sprockets they achieve a near horizontal position over the trough and a high pressure spray is directed downward to remove debris. The trays are then guided to a vertical position as they descend below the operating floor level.

Although not a true "down spray", another modified design (used

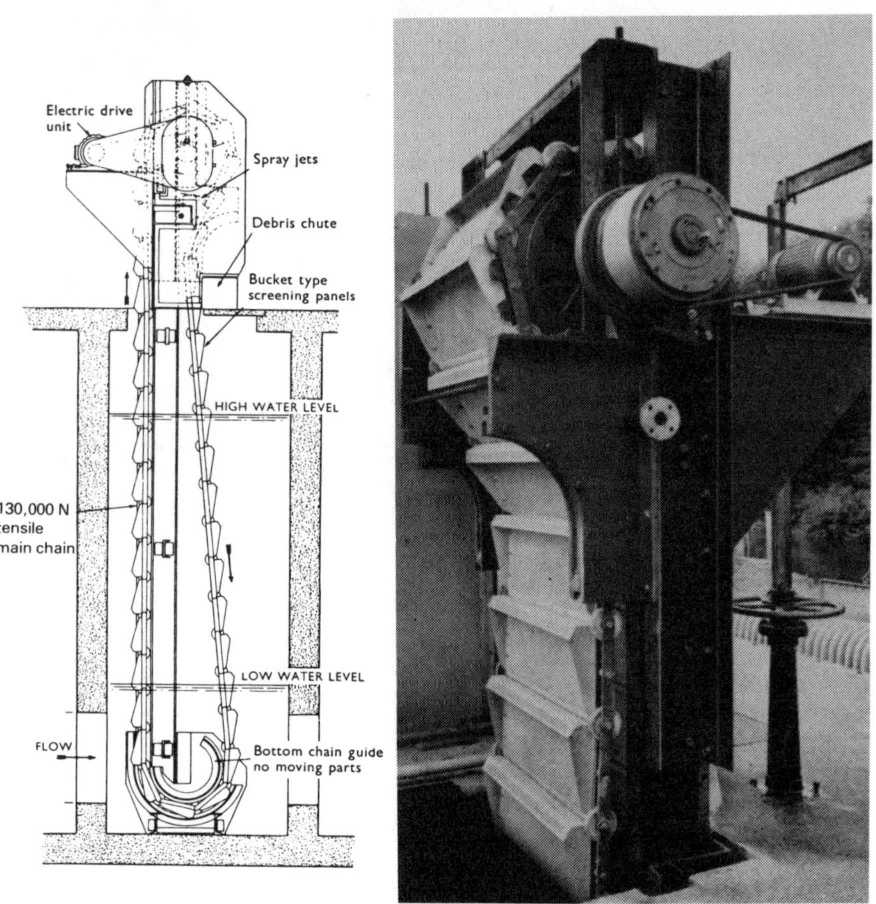

Modified Down Spray Water Screen
Courtesy of William Green, Ltd.

only on Thru Flow screens) accomplishes a similar result with a single headshaft and a guide track arrangement. As the descending trays round the headshaft, they are guided into an angled position over the trough at the discharge point.

Both Down Spray designs provide excellent debris removal. However, the use of guide tracks, extra sprockets and bearings may result in some increased operating and maintenance costs.

Inclined Screen. Inclined Screens are a variation of the Thru Flow Traveling Water Screen that are used primarily to enhance a screens debris carrying capacity.

Inclined Screens are mounted in an intake channel at an inclination of 10 to 15 degrees from vertical, in the direction of the flow (Some screen designs may allow inclination of up to 45 degrees). This angle is often accompanied by an 8 degree inclination of the wire mesh within the basket frame, resulting in a total mesh inclination of 18 to 23 degrees from vertical. The placement of the screening surface at this angle increases the debris retention ability of the screen, and helps prevent large solids from falling off the tray shelf as the trays are lifted out of the flow.

The use of an Inclined Traveling Water Screen should be considered where very large debris loads are regularly expected, or where the size and type of debris makes debris retention on vertical screens difficult. If regular debris loads in excess 15 cubic feet per hour per foot of screen width are anticipated, an inclined screen may be required. Inclined Screens are used most frequently for the screening of woodyard log flumes at paper mills.

The inclination of the screen framework requires a rear debris discharge. The discharge point is located on the downstream side of the screen floor opening, and is slightly cantilevered over the debris trough. This lessens the possibility of debris falling between the trough and the descending trays, into the screened water.

Inclined screens may experience increased chain wear and somewhat higher operating and maintenance costs due to the chain rollers operation in the inclined guide tracks.

Belt Screens A Belt Screen is a simplified version of a Thru Flow Traveling Water Screens. Belt Screens can be used in light duty screening applications with channel widths to six feet, and channel depths to twenty feet.

Belt Screen
Courtesy of Farm Pump & Irrigation Co.

Vertical Belt Screen
Courtesy of Jeffrey Division, Dresser Industries

Belt Screens consist of an endless wire mesh belt in place of the individual screen baskets found in conventional Traveling Water Screens. The belt may operate over a rubber lagged head pulley and slatted foot pulley, or it may be chain mounted and operate over head and footsprockets. They may be installed vertically, or at inclinations of up to 45 degrees.

Belt cleaning is accomplished by a high pressure spray wash system, similar to that used on conventional Traveling Water Screens. To increased the screens lifting capacity and assist in debris removal, auxiliary lifting shelves can be attached to the belt at regular intervals.

Rear View of an Inclined Belt Screen
Courtesy of Pro-Ent, Inc.

At least one manufacturer offers a Belt Screen that utilizes a series of side-by-side, 4" wide mesh belts as the screening surface. These belts are uniformly and automatically twisted 180 degrees after passing the spray wash system to present a clean surface to the incoming flow. The screen can also be fitted with an independently operated chain and scraper unit to aid in the removal of large, or heavy debris.

Platform Mounted. A Double-Entry Dual Flow screen can be mounted from a platform or pier that is supported by pilings, eliminating the need for a concrete channel or well.

The flow pattern of this screen is similar to other Double-Entry screens: raw water enters the screen through the ascending and descending trays and exits through an opening in the center of the screen structure. The screen exit opening is connected directly to the pump suction by ductwork.

Platform Mounted Power Plant Intake
Courtesy of FMC Corporation

This configuration may result in considerable intake construction cost reductions, and the openness of the design allows relatively easy escape for marine life.

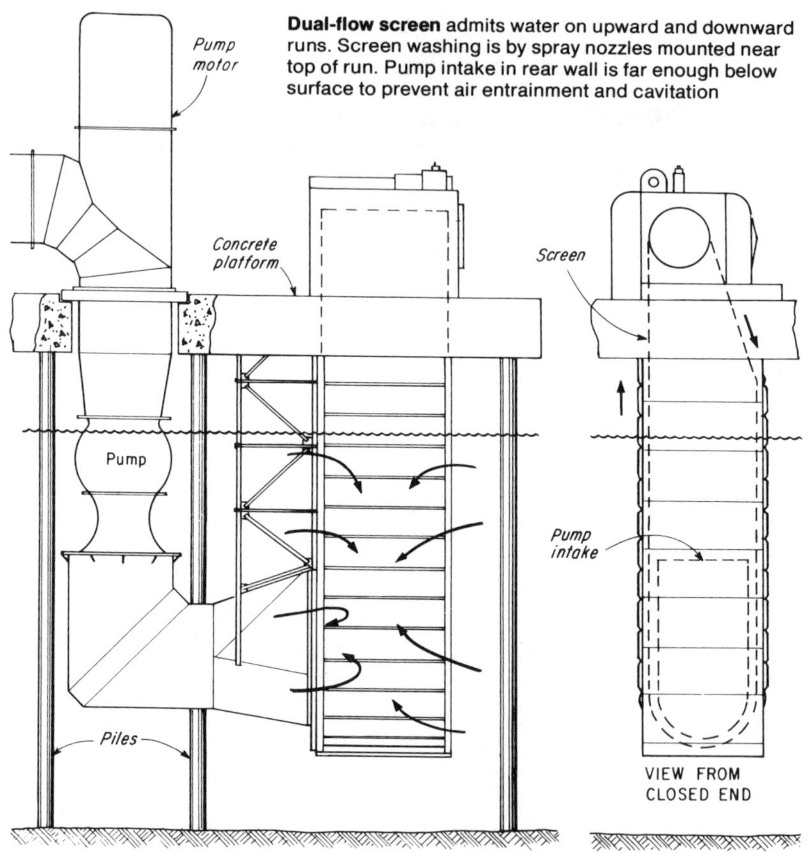

Platform Mounted Screen
Courtesy of FMC Corporation

SCREEN COMPONENTS

Trays (Baskets)

Traveling Water Screen trays, or baskets, are the individual structural frames within which the screen mesh is mounted. The endurance and screening efficiency of a Traveling Water Screen is largely dependent upon its tray design and construction.

Trays may range from 2'-0" to 14'-0" wide, with 10'-0" considered the industry standard for Thru Flow screens and approximately 6'-0" considered the standard for Dual Flow screens.

Trays should be designed to maximize the available screening area and provide sufficient structural strength to withstand the loads imposed by a differential headloss of up to 5'-0". The horizontal tray frame members act as beams to support the loads acting upon the trays, and are designed to provide a seal with adjacent trays to prevent debris from passing between them. This seal point is located on the centerline of the screen chain to insure that a constant clearance of 1/8" to 3/16" is maintained between adjacent trays, even as they are rotated through the boot section.

Some manufacturers offer tray designs that utilize neoprene sealing strips between adjacent trays to provide positive protection when small mesh openings are required.

The lower, or trailing horizontal frame member also serves as a debris lifting shelf. This shelf assists in the retention of debris that does not cling to the screening media.

Wire mesh screening material is assembled to the tray frame using clamp bars and carriage bolts. Vertical supports, spaced at 18" to 36" centers, are located on the downstream side of the mesh to assist in its support and minimize deflection. Most manufacturers install the mesh in the tray frame at a 5-15 degree inclination to aid in the retention of debris on the ascending tray run.

Typical Screen Tray, or Basket Assembly
Courtesy of FMC Corporation

Concave Tray Design
Courtesy of Passavant-Werke

Angle and Bar Tray Construction. Several manufacturers offer tray designs that are constructed of fabricated or commercial steel angles. These horizontal angles may be 1/4" or 3/8" thick, and are jig-welded to end plates to form the frame module. Some Thru Flow designs also utilize curved seal plates that fit between the tray end plates and the chain to insure a seal in the boot area.

Die-Formed Tray Construction. This tray design has horizontal frame members constructed of die-formed steel plate, typically 1/4" thick. These members form overlapping "lips" that provide a positive seal even as the trays flex around the sprockets. The lips are jig-welded to curved end plates that seal the sides of the trays through the radius of the boot section.

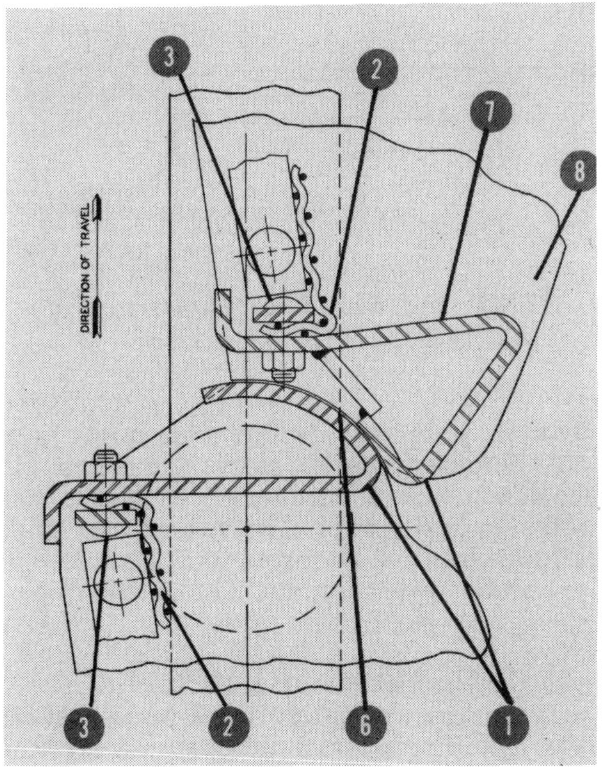

Section View of Die-formed Tray Frame
Courtesy of Envirex, Inc.

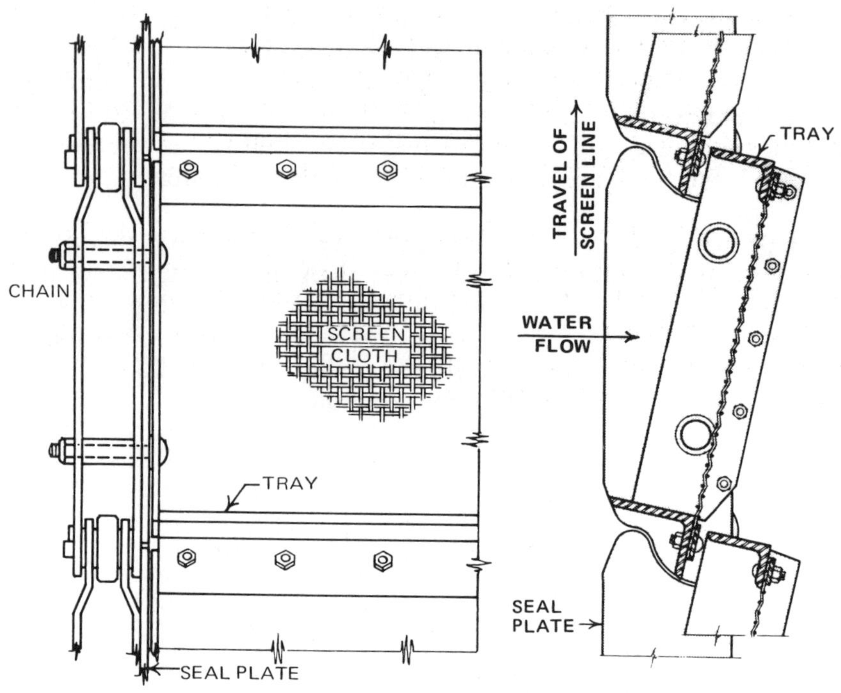

Angle and Bar Tray Construction
Courtesy of FMC Corporation

Fiberglass Trays. Some manufacturers are now offering trays of fiberglass and fiberglass/stainless steel composites. The obvious advantages include greatly reduced weight (33% less than steel) and increased corrosion resistance. The reduced weight allows for easier installation/removal, faster tray travel speeds, and significantly reduces wear on the chain, bearings and other components.

The strength, rigidity, and durability of the fiberglass trays currently available may vary dramatically among manufacturers. Manufacturers should be consulted regarding the specific performance characteristics of the fiberglass trays, and their ability to withstand differential headloss and the flexural stresses encountered in screening operation.

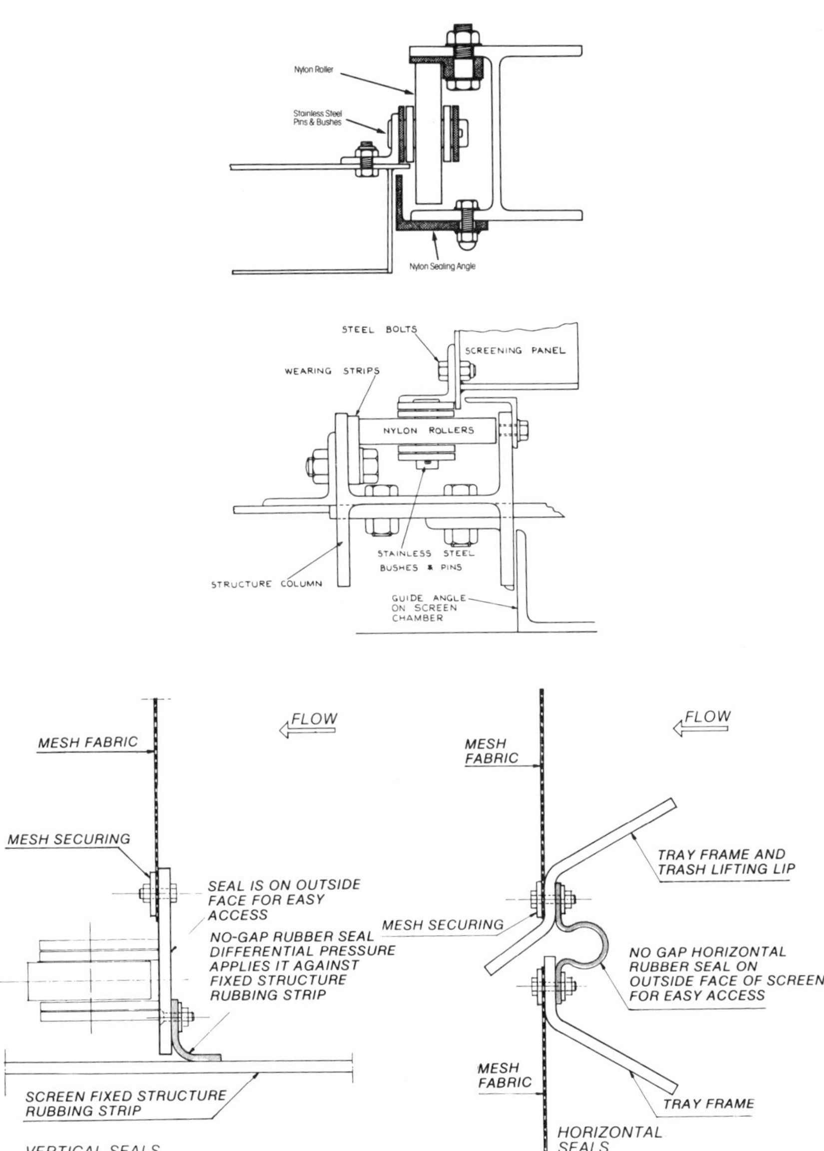

Seals used with Fine Mesh Screen Trays

Top & Center, Courtesy of William Green, Ltd.
Bottom, Courtesy of Beaudrey Corporation

Tray Attachment. There are several methods that may be used to attach trays to the screen chain. Most methods are proprietary to individual manufacturers and include:

* *bolts passing through the chain sidebars and tray end plates*
* *bolting end plates to attachments welded on chain sidebars*
* *bolting tray "shelf" plates to a chain attachment pin*

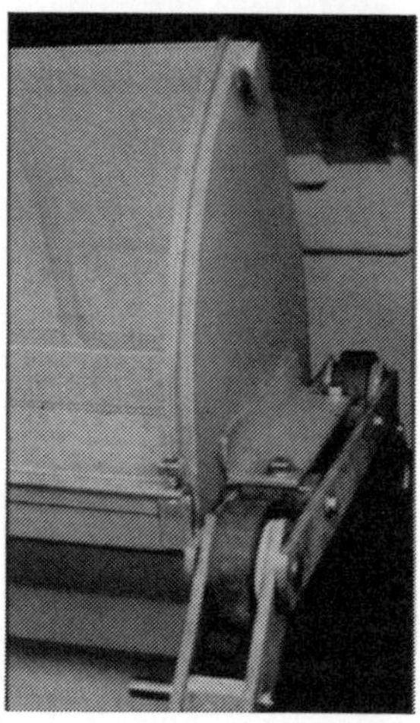

Involute Shaped Tray and Chain Assembly
Courtesy of Davy Bamag GmbH

Auxiliary Backup Beam. If high differential headlosses are anticipated, or if tray widths are greater than 10'-0", a center backup beam support may be considered. This system utilizes a vertical beam, with a stainless steel wear pad, located immediately behind the tray and tied into the main structural framework of the screen. This beam serves to support the tray frames as the deflect under the loads of a water differential. This system is most effective if the individual tray are also equipped with a wear shoe. The wear shoe, manufactured of a low-friction material, will ride up the beam, minimizing tray deflection.

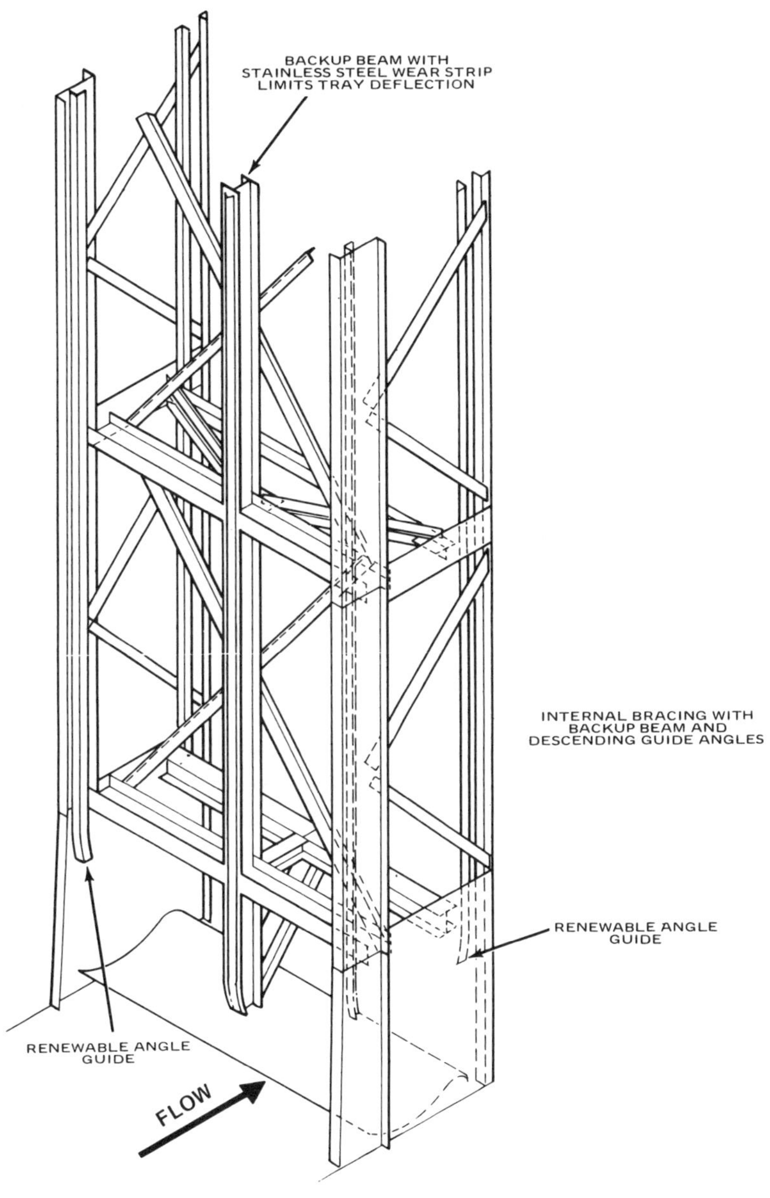

Backup Beam Arrangement
Courtesy of FMC Corporation

(NOTE: The use of a backup beam is intended to provide an added degree of safety when unusual operating conditions occur. Its use is not recommended for normal operation. The screen manufacturer should be consulted for tray widths greater than 10'-0".)

Auxiliary Lifting Lips. Several auxiliary lifting lips are available to solve special debris handling problems. Manufacturers offer auxiliary lip designs that can double or triple the debris handling capability of each tray. These lips may be toothed or plain, curved or flat, and may be bolted or welded to the existing lower debris shelf. For extreme debris loading requirements, a second auxiliary lifting lip can be mounted to the center of each tray.

Mesh

Although perforated plate and expanded metal have been used, woven wire mesh is the most common screening media in use on Traveling Water Screens. It is available in a variety of wire diameters and weaves, with square or rectangular openings of virtually any size, and can be manufactured to suit most material requirements. A typical Traveling Water Screen uses a plain or double crimped weave, with a 3/8" square opening of #14 gage W&M (0.080"diameter) type 304 stainless steel wire.

The typical range of mesh openings is 1/8" to 3/4", although some installations have used mesh with openings as small as 0.1mm. If a fine (<1/4") mesh is required, careful consideration should be given to the water velocity through the wire. Even if a fine mesh screen is designed within the

normal (1 to 2.5 fps) velocity range, it should be noted that the finer openings will retain more debris, more quickly, which may lead to a rapid buildup in headloss across the screen. A fine mesh may be impractical with many Traveling Water Screens because their tray and framework designs are not capable of maintaining a seal less than 3/16" between adjacent trays, and the ends of the trays and the boot section and/or framework.

The individual strands of wire that make up the screen mesh generally have the smallest cross sectional area of any part of the

Traveling Water Screen. It is important that they are as thin as practical so that the maximum open area of the mesh can be realized. However, if the wire strands are too thin they may be vulnerable to corrosion, damage or puncture. The use of stainless steel or monel wire material is recommended over carbon steel, or galvanized carbon steel in most applications.

An increasing number of installations are using a synthetic polyester monofilament screening fabric. This material is relatively rugged, corrosion resistant, inexpensive and is much lighter than stainless steel wire.

Rectangular openings. Although square woven mesh is most common, the use of a rectangular or oblong mesh opening may have advantages in some applications. A rectangular mesh opening of 1/8" by 1/2" has an open area greater than a mesh with a 1/4" square opening of the same diameter wire. In addition, debris removal efficiency of the 1/8" by 1/2" weave is usually much higher.

Wedgewire screen. Wedgewire screen panels have been used in lieu of woven wire mesh successfully in some screen installations. The added weight of most wedgewire screens should be considered prior to their use.

Interchangeable mesh panels. Trays can be furnished with quick-change mesh inserts that allow the operator to replace the standard mesh with one better suited for seasonal operating requirements.

For example, fine mesh screen inserts may be used to prevent the entrainment of fish eggs and larvae during the summer breeding season. After the breeding season, the original larger mesh screen inserts can be re-installed.

Tray Chain

The chain that carries the tray assemblies is a heavy roller chain. Thru Flow screens are usually provided with a 24" pitch chain. Dual Flow units generally utilize 18" pitch chain, although 12", 19", 21", and 24" pitch chain is available.

Standard duty screen sidebars are 3/8" thick and 2-1/2" to 3" wide, fabricated from carbon or stainless steel. Heavy duty, or very deep screens may have 1/2" thick sidebars. European designs generally incorporate straight sidebar chain and United States manufacturers offer chains with double offset chain sidebars.

Chains are usually furnished with grease lubricated carrier roller joints having heat treated pins, rollers, and bushings. Although some fresh water installations use carbon steel materials for chain joints, the use of a heat treated 400 series stainless is most common for both fresh water and seawater applications. Problem installations may consider using a 17-4 precipitated hardened (PH) stainless steel or monel chain joints. Non-metallic molded nylon chain rollers are offered as a standard by several European manufacturers, and are now available from some United States manufacturers.

Non-lubricated chain designs are available from most manufacturers. These chains usually incorporate a cast chrome or molded nylon roller and 17-4 PH pins and bushings. The use of non-lubricated chain should be evaluated on a case-by-case basis.

Some manufacturers offer chain links equipped with a glide block arrangement. A replaceable glide, or wear shoe can be mounted to the outside sidebar of each chain link to minimize lateral movement and assure that the chain tracks properly.

Footshaft/Footsprocket

Dual Flow Foot Terminal. Most Dual Flow screens are equipped with a "roll-around" foot terminal in place of the conventional footshaft/footsprocket arrangement. The screen chain is guided in a 180 degree arc through the foot terminal, or boot section, by a formed steel, or cast iron guide rail. This design eliminates any permanently submerged moving parts.

Thru Flow Footshaft. Thru Flow screens are equipped with a pair of footsprockets, or footwheels mounted on a shaft in the boot section of the screen frame. The maintenance and inspection of the footsprockets is difficult and costly and their proper operation is critical to the performance and reliability of the screen.

There are two shaft configurations offered; a "live" shaft that revolves within two bearing blocks and is fitted with fixed sprockets, and a fixed "dead" shaft with bushed sprockets revolving on it. The bushing material for live shaft designs is either bronze or a graphite impregnated, molded canvas laminate, and the bearing block for standard dead shaft designs is usually manufactured of a self lubricating bronze material.

Applications that require continuous operation, or anticipate high levels of silt or grit should consider an alternate material selection for the footshaft/footsprocket bearing material. An alternative that has been used successfully in such applications is a centrifugally cast chrome-nickel-boron alloy bushing & sleeve arrangement. This design utilizes a cast bushing operating over a similar sleeve that has been press fit onto shaft. This material is considerably more expensive that the previously mentioned "standard" materials, but the expense can usually be justified.

A split shaft arrangement will reduce labor costs when the inevitable repair or replacement of footshaft components is required. The footshaft may be fabricated in two or three pieces, each held together with ribbed compression couplings. If removal of the footshaft is required, it can first be disassembled within the screen, and removed in sections.

"Roll-Around" Foot Terminal
Courtesy of FMC Corporation

Headshaft Assembly

Like other drive components, the headshaft must be adequately sized to withstand the NEMA rated stall torque of the motor. The headshaft diameter is sized based on the combined torsion and bending moments and the allowable torsional deflection. Headshafts are available of high carbon solid steel shafting material in diameters ranging from 3-7/16" to 6-1/4". Some manufacturers offer torque tube headshaft arrangements to insure that overload conditions do not result in torsional deflection and bending that could twist or skew the trays.

Standard headsprockets for 24" pitch chain have six teeth and a pitch diameter of 48". Space limitations on some existing and replacement screens, (and a very few new installations) may require a five tooth sprocket design, with a smaller pitch diameter.

Headsprockets should be provided with replaceable tooth inserts of hardened stainless steel, fitted to the sprocket teeth. Replaceable tooth pockets are also offered by some manufacturers.

Segmented sprocket designs are offered as options by most manufacturers to facilitate repair and/or removal of the headsprocket.

The centerline of the headshaft is typically located 3'-0" to 4'-6" above the operating floor level.

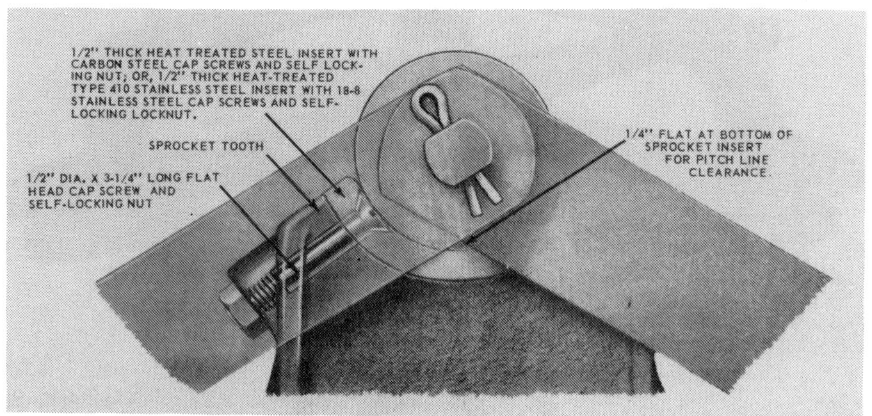

Headsprocket Tooth Insert
Courtesy of Envirex, Inc.

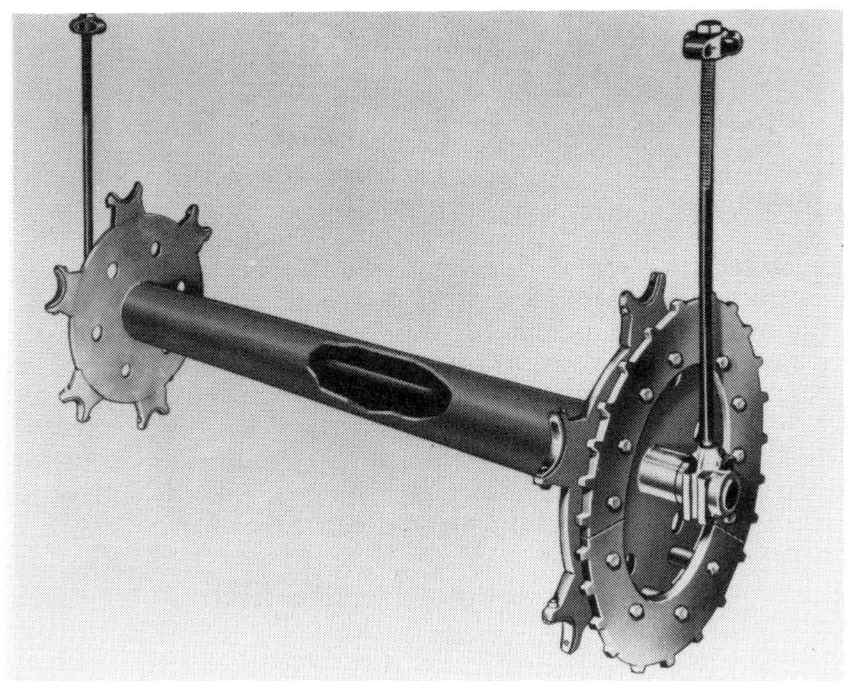

Head shaft — Inside Drive with Torque Tube

Symbol	Description
A	Shaft
B	Key
C	Sprocket (with bosses)
D	Sprocket (plain)
E	Tooth insert
F	Bolts w/locknuts
G	Driven sprocket
H	Bolts w/locknuts
J	Takeup bearing
K	Takeup bearing bushing
L	Takeup screw
M	Thrust bearing
N	Adjusting nut
P	Jam nut
Q	Torque Tube

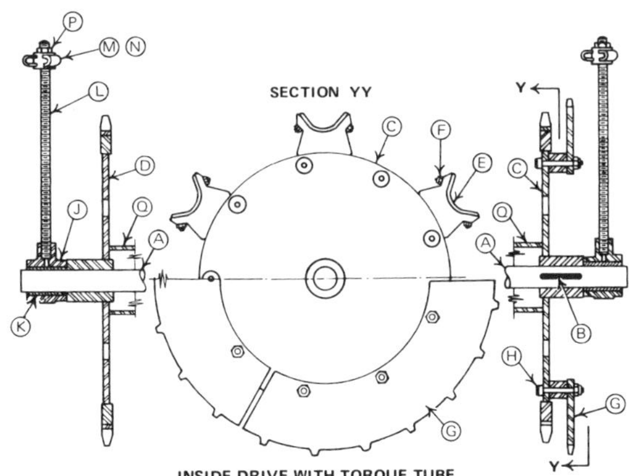

Torque Tube Headshaft Arrangement

Courtesy of FMC Corporation

Takeup Assembly

The takeup assembly provides for adjustment of the tension of the tray carrying chain by allowing the headshaft assembly to be raised or lowered.

The headshaft assembly rotates in two adjustable bearing blocks fixed to threaded, stainless steel or bronze takeup screws. The upper ends of the takeup screws are equipped with adjusting capstans and roller type thrust bearings. The takeup screw is supported by the screen headframe.

The standard Traveling Water Screen takeup bearings are of the grease lubricated, bronze bushed type, although most manufacturers offer optional sealed, antifriction type takeup bearings.

Inadequate chain tension is one of the primary causes of premature screen wear. Both over- and under-tensioning of screen chain are common operational problems. For this reason, most manufacturers offer equipment to assist or automate the chain tensioning procedure. Devices available include load cells with chain tension indicators, hydraulic tensioning systems, and spring-loaded adjustment equipment.

Framework

The main framework of a Traveling Water Screen is constructed of formed plate and structural shapes to provide free-standing support for the screen operating mechanism, and chain guide tracks for the revolving trays. Vertical flanges on each side of the frame mate with guideways embedded in, or bolted to the channel walls to position the screen assembly in the channel and prevent the passage of debris around the unit.

Traveling Water Screens may be supported at the operating floor or the channel invert elevation. If the screen is supported at the operating floor level, the headsection should be equipped with jacking, or leveling screws. The jacking screws should be located over bearing plates that are fixed to the operating floor to allow for final screen adjustment after installation.

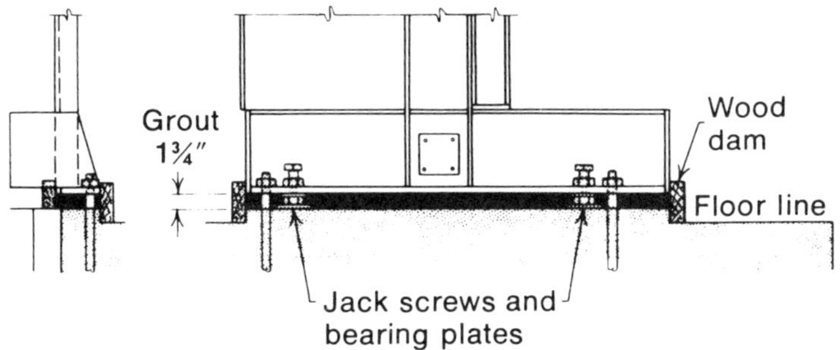

Side View of Headsection Leveling Arrangement
Courtesy of FMC Corporation

If the screen is supported at the channel invert elevation, the boot section merely rests on the bottom of the channel, with shims placed beneath it to level the screen shaft assemblies.

The screen framework is made up of several frame sections that are bolted together with splice plates for easy handling and erection in the screen channel. These standard sections include the headsection or head terminal, upper intermediate section, intermediate section, and the boot section.

The headsection is that portion of the screen framework that is located above the operating floor level. The headsection supports the headshaft assembly and spray system, and may be equipped with a machinery platform to support the drive unit. The headsection is usually fabricated of a minimum 1/4" thick steel, and should include provisions for attaching a hoist and lifting the entire screen assembly from the well. Headsections may also include optional jacking pads. These pads are welded perpendicular to headsection sideframe and provide a convenient location to insert an hydraulic jack to break the screen loose from the well and assist in its removal.

Access to the inside of the area of the headsection to inspect or maintain the spray system and headsprocket tooth inserts is usually very limited. Most screen designs require removal of the splash housing and one or more trays. Access can be greatly facilitated by the addition of a access opening in the side panel of the headsection framework. This opening, in conjunction with an access platform spanning the width of the screen between the ascending and descending trays, will provide passage and a suitable working platform.

Screen Headsection
Courtesy of Hawker Siddeley Brackett

The headsection is bolted to an upper intermediate frame section at approximately the operating floor level. This frame section joins the headsection to the remaining frame sections, and usually varies from 5 to 14 feet in height.

The intermediate framework is constructed in one or more standard sections measuring approximately 10'-0" high. These sections are usually fabricated of 3/8" thick plate and are well stiffened and cross braced, forming a rigid box-like frame. The sideframes of the intermediate sections provide vertical guide tracks for the tray chain.

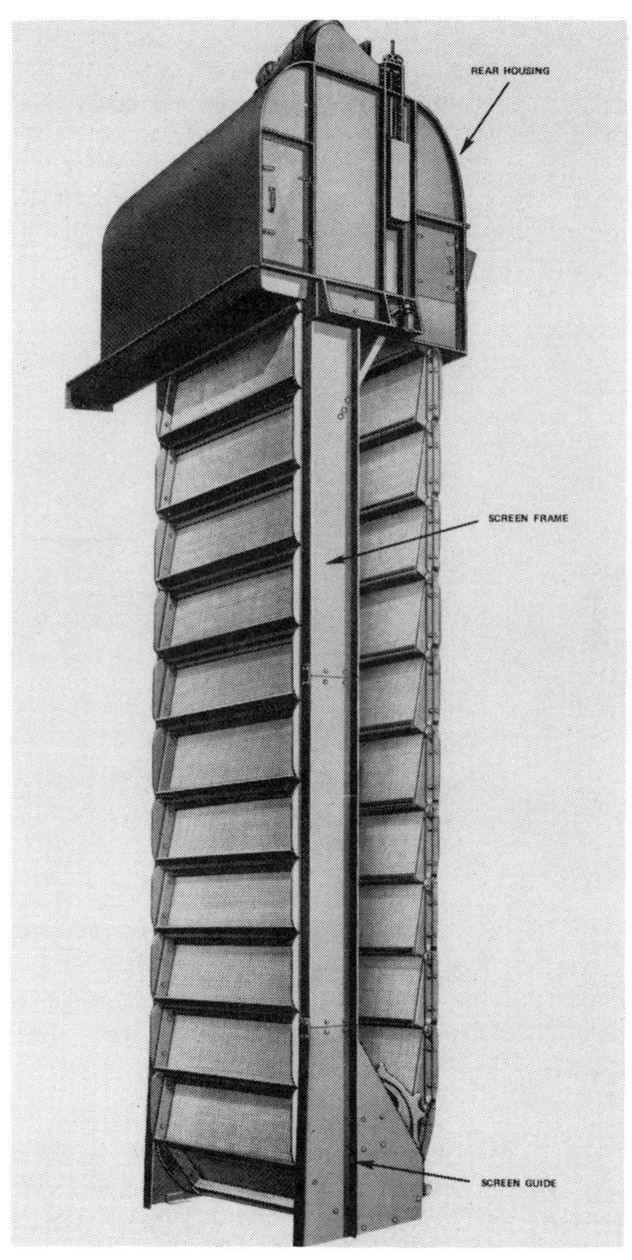

Two Post Screen Assembly
Courtesy of FMC Corporation

Continuous vertical flanges on the outside of each sideframe are designed to fit into the guideways set in the concrete channel walls to prevent debris from passing around the screen frame. The intermediate sections should be constructed with provisions to allow the insertion of stop beams to support the frame at intermediate levels during screen installation, maintenance or removal.

Dual Flow Screen Frame
Courtesy of Hawker Siddeley Brackett

Frames of Double-entry Dual Flow screens may be equipped with an automatic slide gate that can be activated during periods of extreme differential headloss. This gate is located on the screen side frame and, when activated, allows water to bypass the trays and enter the center of the screen, relieving an extreme headloss condition.

Thru Flow screens are available with a "two" or "four" post frame design. Two post frames are designed to provide chain guide tracks for the upstream, ascending trays only. For deep installations, or other applications where it may be necessary to also guide or confine the descending trays, a four post frame with "downrun" guide angles can be provided. Dual Flow screens and screens with tray widths in excess of 10'-0" usually require four post frame construction.

The boot section is the bottom-most frame section. The footshaft assembly is located in the boot section, with its centerline approximately 2'-6" from the channel bottom. The boot section is equipped with a curved bootplate and track guide arrangement to seal the area under the screen as the descending trays revolve around the footsprockets and begin their ascent. Most Dual Flow screens use a curved guide rail arrangement that eliminates the use of the footshaft/footsprocket and bootplate.

For applications that experience high levels of sand or silt, the boot section may be equipped with a baffle plate that extends from the channel bottom to a point level with the footshaft. This baffle will assist in keeping abrasive sand or silt in suspension, and prevent it from settling in the boot area.

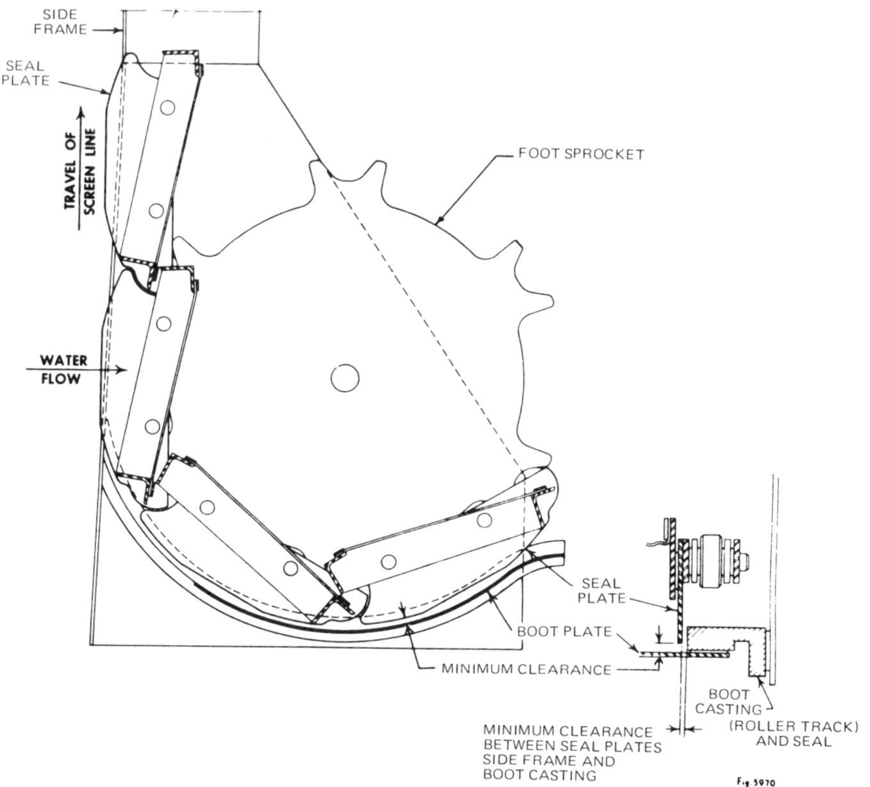

SEAL PLATE ACTION IN FOOT SECTION

Courtesy of FMC Corporation

Shop Assembled Screen
Courtesy of FMC Corporation

The chain guide tracks are usually the only portion of the screen framework that is subject to wear under normal operating conditions. The life of the framework may be increased by utilizing bolted-type replaceable track angles, or lining the wear surface of the angles with replaceable steel or nylon wearing strips.

Redwood, cypress, nylon, or neoprene sealing strips may be used to further seal the area between the framework and the ends of the trays. These sealing strips can bolted to the upstream flanges of the mainframe to engage the tray endplates.

Flexible sealing strips can also be installed to seal the boot section of a Thru Flow screen.

Frameless Designs. Some Dual Flow designs are provided without a conventional main frame or superstructure. These designs utilize vertical guide channels secured to the walls of the intake chamber that serve as chain tracks.

This arrangement maximizes the use of the channel width, but does not allow the screen to be removed as a unit for inspection or repair.

Spray Wash System

Debris is removed from the screen mesh by a system of high pressure water sprays directed through the trays into a debris trough. Most Thru Flow and Dual Flow Traveling Water Screens use a front cleaning design that removes debris from the ascending trays on the upstream side of the screen.

Proper operation and maintenance of the spray system is essential to satisfactory screen performance. If a Traveling Water Screen spray system is not operating properly, the screen simply becomes an expensive debris conveyor.

Spray water requirements vary based upon the screen width, and the type and amount of debris. A typical spray wash system consists of a single row of venturi type spray nozzles bolted or threaded to a spray pipe to provide a complete, and slightly overlapping spray coverage of the baskets. If debris loading is very heavy, or debris is particularly tenacious, a second spray pipe and nozzle arrangement may be required.

Typical Spray Wash System

Dual Flow screen manufacturers usually recommend operating the screen spray system at approximately 35 psi, and 18 gpm per foot of tray width.

Thru Flow screen manufacturers generally recommend operation of the spray system within a discharge range of 20 to 30 gpm per foot of tray width at a pressure range of 60 to 100 psi, respectively. Field tests indicate that pressures of 60 psi are adequate for most Thru Flow applications.

The higher wash water pressure is recommended on Thru Flow screens because it is necessary to insure positive debris removal to prevent carry-over problems. The design of a Dual Flow screen has an inherent protection against carry-over, and the lower spray water pressure has proven satisfactory in most applications.

Excessive water pressure can also result in operational problems. Spray water pressures greater that 100 psi can result in a cascading effect when turbulence caused by the spray water hitting the inside of the splash housing results in debris being splashed or blown up onto clean trays.

Most spray systems include an automatically operated valve that is actuated before the screen trays are revolved. Some systems are also equipped with a pressure sensing switch that prevents the screen from being rotated until adequate pressure is developed in the spray water line. The system should also be furnished with an automatic drain system to prevent damage from freezing when the screen is not in operation.

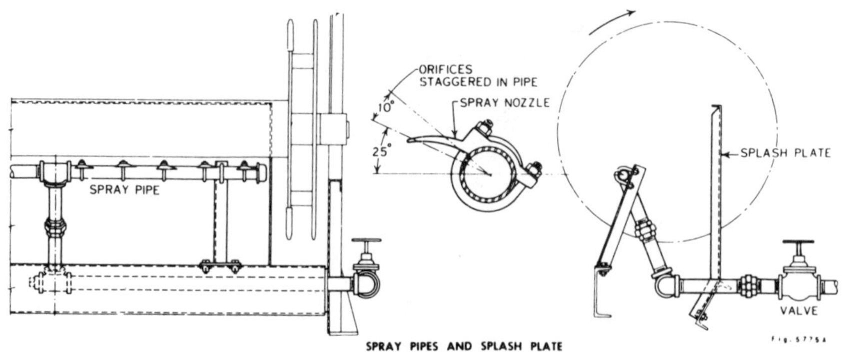

SPRAY PIPES AND SPLASH PLATE

Courtesy of FMC Corporation

Spray wash water is usually obtained from the screened water on the downstream side of the Traveling Water Screen. The use of an in-line, self-cleaning wash water strainer is recommended to minimize clogging of the spray nozzles. It is recommended that the strainer be equipped with filter openings that are 1/3 the size of a the nozzle orifice.

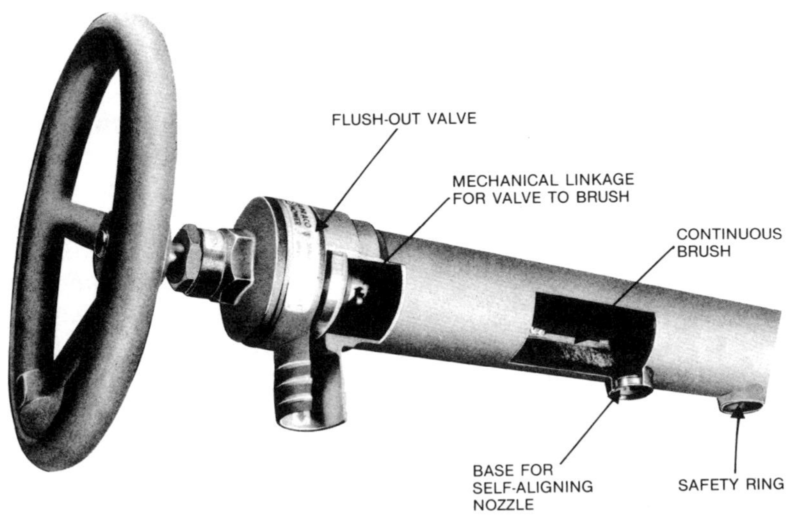

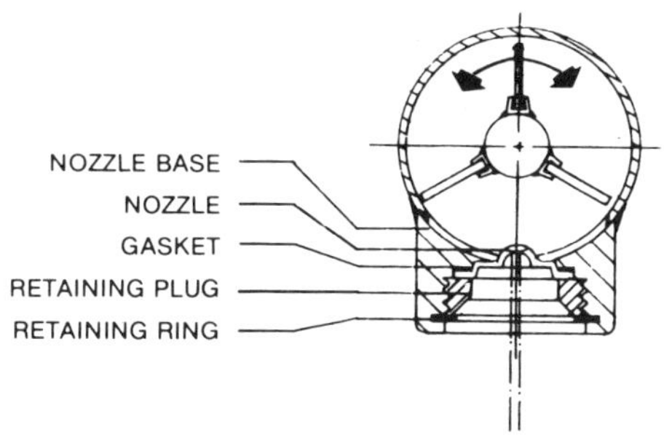

*Nozzle tip protrudes into pipe, where brush cleans the orifice **and** the pipe when turned.*

Spray Header with Internal Brush Cleaner
Courtesy of Spraco, Inc.

A spray header and nozzle cleaning device may be provided with the screen spray system. One cleaning device consists of bristles mounted on a long rod located inside the spray header and connected to an external handwheel. The brushes can be manually rotated, by means of a handwheel, to periodically remove accumulated solids.

Splash Housing

Splash housings enclose the screen headsection and debris trough, contain the wash water spray, and direct it into the debris trough. Although semi-enclosed splash housings are available, totally enclosed housings are normally furnished for safety reasons.

Housings are usually constructed of molded fiberglass or light gauge stainless steel. The portion of the housing (usually the "front" housing) that directly covers the spray system and debris trough should be watertight with appropriate seals and gaskets to prevent leakage. The front housing must have sufficient flexural strength to withstand the constant pressure of the spray water. Dependent on the materials of construction, the front housing should meet the following minimum specifications:

Carbon Steel ....#10 gauge (0.1345") sheet

Stainless Steel......#12 gauge (0.1046") sheet

Fiberglass...........3/16" thick, molded
13,000 psi avg tensile strength
25,000 psi avg flexural strength

The rear housing is provided for safety purposes and to assist in the containment of wash water overspray. It should be easily removeable, or have a suitable opening to provide easy access to the trays and chain for inspection and maintenance purposes.

All housing hardware and fasteners should be manufactured of stainless steel.

Debris Trough

The debris trough collects the spray wash water and refuse removed from the screen, and transports away from the intake area. Debris troughs should have a minimum depth of 6", a minimum width of 1'-2", and a minimum slope of 1/8" in 12" to assure proper flow in the trough. Multi-screen installations may use a common trough that spans the width of the intake. The portion of the trough between adjacent screens should be covered with checkered floor plate or grating.

Debris troughs are usually cast into the concrete intake structure during its construction with the top of the trough located at the operating floor, or deck elevation. Some installations utilize above-ground troughs of steel, stainless steel or fiberglass construction.

The edge of the debris trough is usually located 4" to 10" from the surface of the screening media. This gap must be bridged with a carry-over prevention deflector plate to prevent water or debris from falling back into the channel between the debris trough and the trays. The design and performance of this deflector is especially important on Thru Flow screens.

A typical deflector consists of a fixed, inclined plate fitted with an adjustable, flexible neoprene edge. The deflector can also be molded as part of the screen housing. Some manufacturers offer a deflector that is designed to follow the contour of the moving trays, to minimize the width of the gap.

Screens that utilize a front cleaned spray system/trough arrangement to clean ascending trays may also utilize a deflector on the descending side of the machine. This secondary deflector can be installed without a spray system, and with or without a trough. The deflector helps prevent tramp debris that was not removed by the front spray system from falling into the downstream side of the screen. For example, horseshoe crabs often cling to the screen mesh and are not removed by the spray system. As they ride over the headsprockets and begin to descend they may be discharged by gravity, onto the secondary deflector.

Drive Unit

Traveling Water Screen drive components include the electric motor, speed reducer, coupling, drive sprocket, driven sprocket, drive chain, and fluid coupling or shear pin overload protection system. Some manufacturers may provide a hollow shaft speed reducer mounted directly to the headshaft in place of the drive/driven sprockets and drive chain arrangement.

Drive unit components are selected to provide the capability of starting and/or operating the Traveling Water Screen under a specified loaded condition without exceeding the NEMA rated starting and stall torque of the motor. The loaded conditions that are usually considered include starting the screen at a specified headloss and/or operating it continuously at second specified headloss, at the maximum tray travel speed. Motor ratings for most applications range from 3/4 to 7-1/2 horsepower.

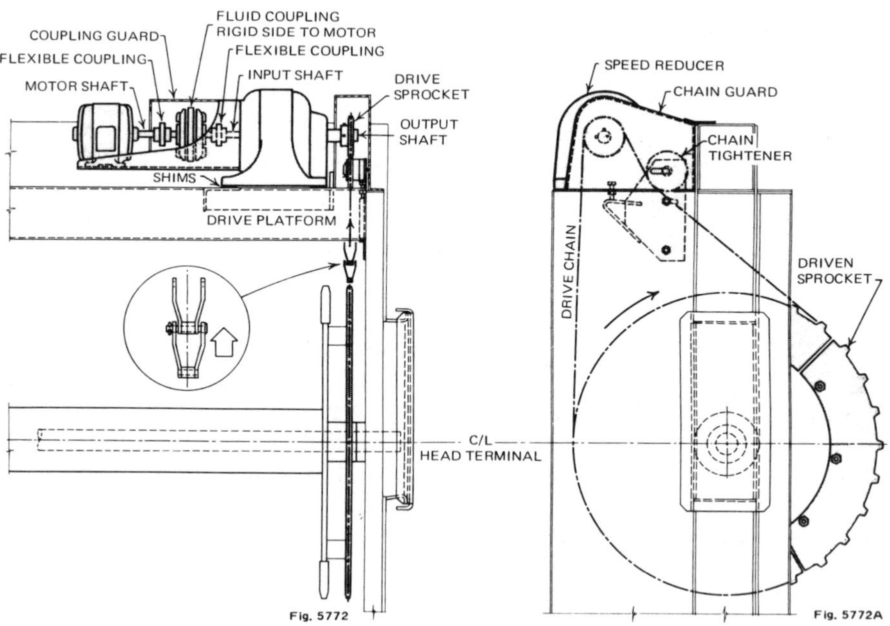

Typical Screen Drive Arrangement
Courtesy of FMC Corporation

Shaft Mounted Drive Unit
Courtesy of Hawker Siddeley Brackett

The screen drive unit should be designed to start the screen under the loads imposed by a minimum 2'-6" differential headloss. Heavier-duty applications may require the screen to start under 5'-0" headloss.

The industry standard tray travel speed is 10 feet per minute (fpm). At this speed, clean trays are introduced to the flow at a rate that can usually keep differential headlosses under control. If heavy debris loading conditions occur frequently, it may be desirable to operate at a higher tray speed. Typical high speed operation is 20 fpm, although tray speeds as high as 45 fpm have been provided.

Installations that are operated continuously, or experience light debris loading, may be operated at lower tray speeds. Reduced travel speeds of 2.5 or 5 fpm will decrease wear and extend the life of screen components in direct proportion to the amount of speed reduction. However, these installations should consider the use of dual or multi-speed drive units that provide the alternative of a faster tray speed for times when debris loading is higher than normal.

Multiple tray speeds may be accomplished by using a two speed or four speed drive motor, or a multiple speed transmission. Variable motor speed operation can be obtained through the use of a variable frequency, variable voltage, motor speed control system.

TRAVELING WATER SCREEN DESIGN & SELECTION

Traveling Water Screens must be designed to perform with a minimum of maintenance under a wide variety of operating conditions. They may be located on a river, reservoir, lake, or ocean, and subject to large fluctuations in flow conditions, debris loading, water depths and salinity.

The information that factors into the determination of the size of a Traveling Water Screens screen for a specific installation includes:

- *Maximum, Average Flow*
- *Maximum, Minimum, Average Water Levels*
- *Number of Screens*
- *Wire Mesh Size*
- *Velocity through Mesh*
- *Tray or Channel Width*
- *Type of Duty or Service*
- *Starting/operating headloss requirements*

As mentioned in the previous chapters, there are two types of Traveling Water Screens that are most common; the "Thru Flow" and the "Dual Flow" designs. Selection procedures for both designs are similar. Unless specifically noted, or sufficiently obvious, it may be assumed that the procedures described herein apply to both types of screens. The selection of the screen model (Dual or Thru Flow) is usually a matter of the engineers or clients preference, although an economic evaluation can determine whether a Dual Flow or Thru Flow design is more cost effective.

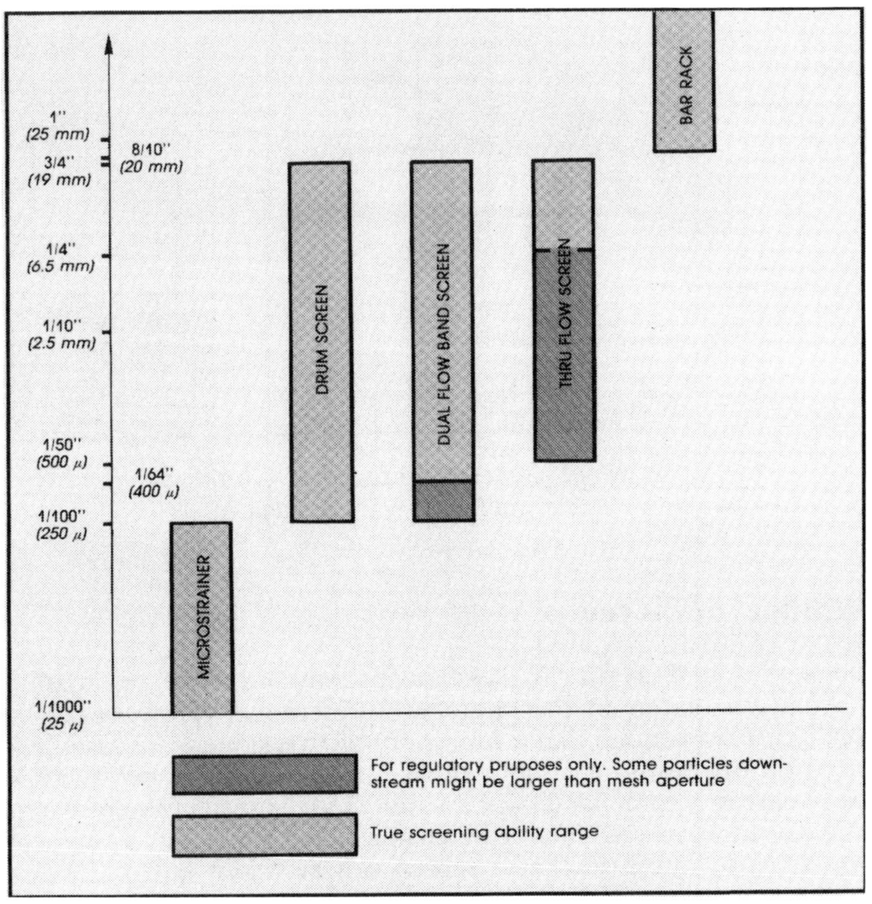

Screening Equipment Options

Courtesy of Beaudrey Corporation

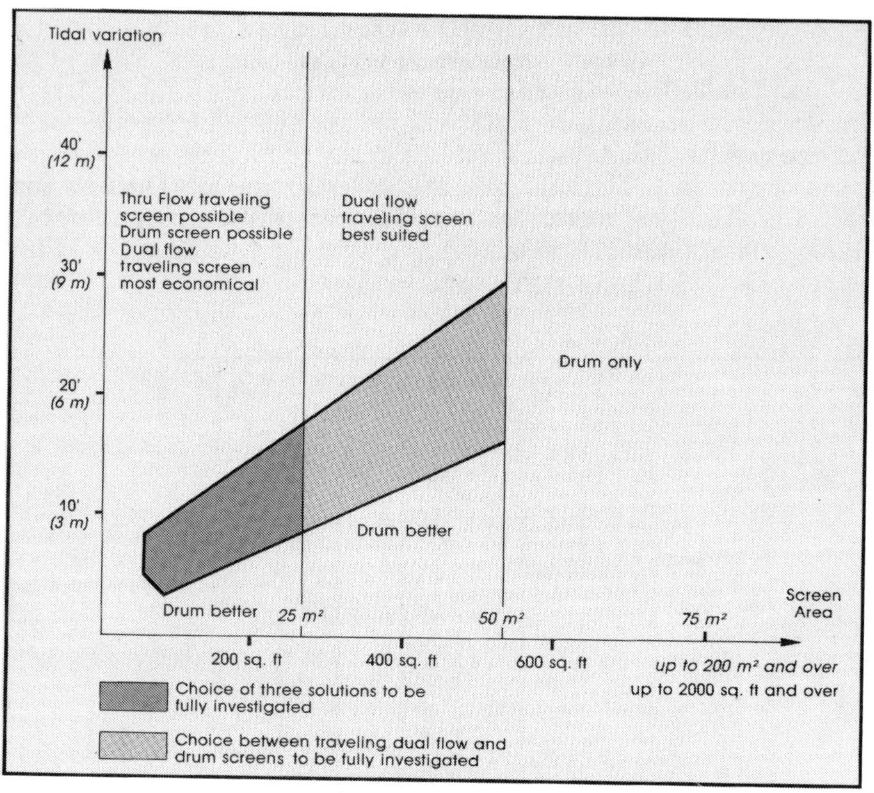

General Screen Selection Chart
Courtesy of Beaudrey Corporation

Number of Screens

The number of Traveling Water Screens required for an installation is usually a matter of the engineers preference. However, it is usually recommended that a minimum of two screens be installed. Based on operating and maintenance considerations, two narrow screens are recommended over a single wide screen. When two or more screens are used, the intake channels and screening equipment should be designed so that a screen can be taken out of service without adversely affecting the operation of the remaining screen(s).

When possible, all screens at a single installation should be of the same model, manufacturer, and size. This will insure interchangeability and reduce the required spare parts inventory.

Future screening requirements must always be considered. If it is anticipated that additional screens may be required later, the size and location of the addition should be considered. It is often cost effective to complete the civil work for subsequent plant additions during initial plant construction.

Intake Channels

Traveling Water Screens are installed side-by-side in concrete channels with each screen, or set of screens separated from the adjacent screens. Screens are usually preceded by a coarse bar rack to prevent large debris from damaging the screen. These bars are spaced to provide 2" to 5" openings.

When selecting a site for an intake, the engineer should first consider the available maximum, minimum, and average(normal) water levels. In most locations these water levels are fixed and very little can be done about them.

It is recommended that the screen channel bottom be at, or above the natural bottom of the water body. If a new channel bottom is dredged, problems due to sedimentation may arise as the artificial recess fills with silt or other sediment when the screen is not in use.

Although the minimum water level is normally the limiting factor in overall screen design (see "Velocities"), the maximum water level is usually considered when determining the screen's sprocket centers. It is necessary to locate the screen headworks at an elevation that will insure that it is not submerged during flood conditions. This requirement may result in a considerable amount of freeboard.

Channel Width

Thru Flow Designs. Thru Flow Traveling Water Screens do not utilize the entire channel width as effective screen area. Ten to twenty percent of the channel area of a 10'-0" wide screen may be restricted by the screens framework and the guideways required to locate and fix the position of the screen. The interface of the "guides" and screen frame also provides a labyrinth seal, preventing debris from bypassing the sides of the screen.

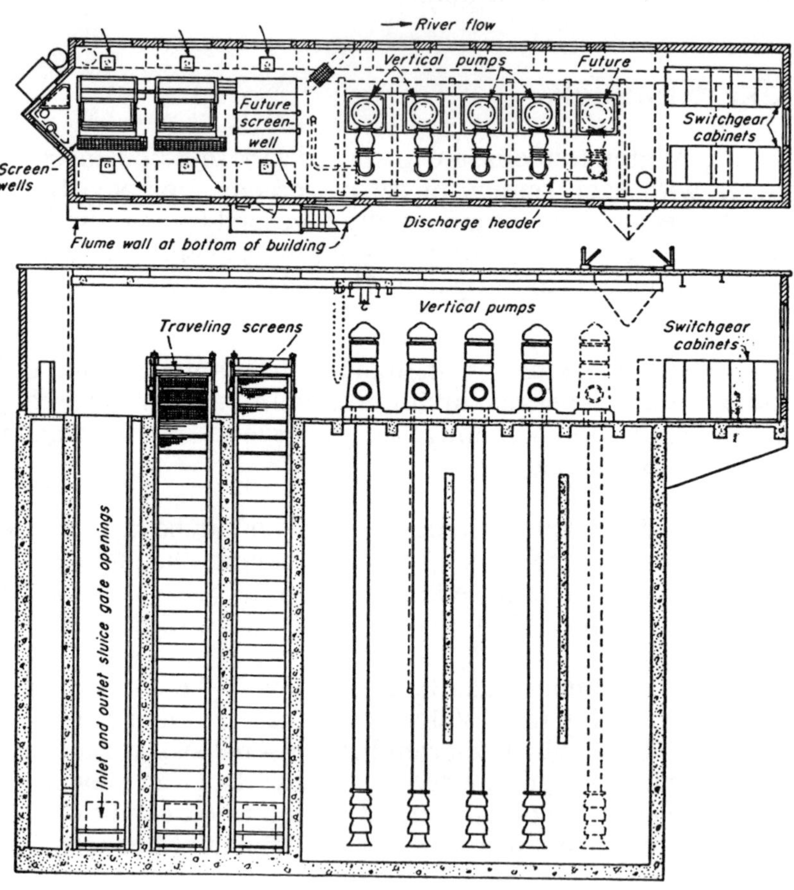

Wet-type River Pumphouse with Vertical Pumps

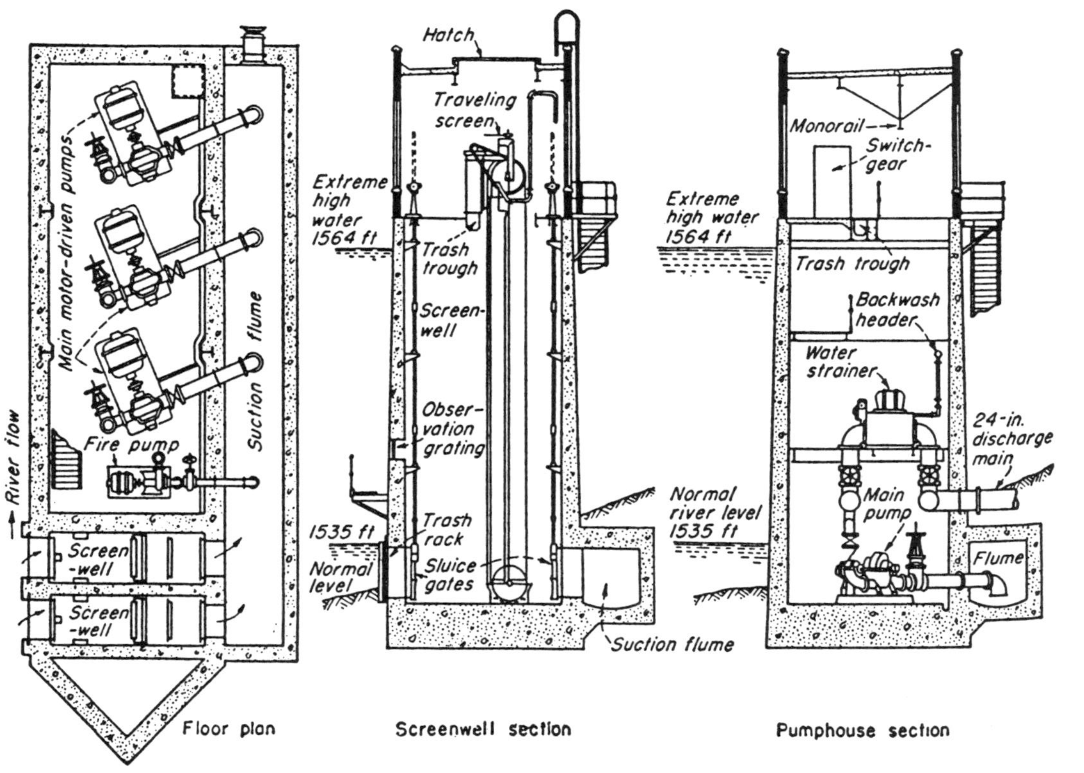

Dry-type River Pumphouse with Horizontal Pumps

The use of vertical guides embedded or recessed in the channel walls is the preferred method of positioning Thru Flow screens. Embedded style guides consist of a "u-shaped" casting grouted into the channel walls so that the legs of the "u" are flush with the face of the wall. The screen is designed with a vertical guide flange as part of its main frame that will fit securely within the pocket of the embedded guide. A screen designed for use with an embedded style guide arrangement typically requires 1'-2" of the overall channel width. Thus, an 11'-2" wide channel will be required for a Traveling Water Screen with a tray width of 10'-0".

Bolted style guides are most often be used where channels are existing, or their use is preferred for other design requirements. A bolted style guide consists of steel angles or a formed plate, bolted directly to the face of the channel walls. The protruding legs of the guide form a pocket to contain the frame guide flange. A bolted style guide arrangement will typically require 1'-8" of the overall channel width.

Dual Flow Designs. In a Dual Flow Traveling Water Screen the effective screen width is oriented in the direction of the flow and has little bearing on the channel width. The minimum channel width required for most Dual Flow Screens is two times the headsprocket pitch diameter.

Because of the variety of headsprocket sizes, screen configurations and hydraulic considerations, it is recommended that manufacturers be contacted for final recommendations.

Velocities

As with the well depth, the width of a Traveling Water Screens is usually determined by considering the extreme operating conditions. This condition usually occurs when the maximum flow occurs during a period of minimum water depth. When these conditions occur simultaneously, the water velocity through the screen increases and the differential headloss across the screen may reach dangerous levels.

In most applications a design velocity range from 1 to 2 fps will result in efficient and satisfactory operation. However, it should be noted that with all things equal, a screen with a velocity of 2.0 fps will be twice as wide as one with a velocity of 1.0 fps.

Some applications may require velocities as low as 0.5 fps, or justify velocities as high as 3.0 fps. The selected design velocity should be based on the consideration of a number of site specific factors:

* *Expected frequency and duration of extreme operating conditions.* If the extreme operating conditions occur infrequently or for very short periods of time, it may be acceptable to operate at velocity ranges of 2.0 to 3.0 fps.

* *Type and quantity of debris to be removed* If debris loads are normally very light, and consist of easy to remove debris, it may be acceptable to operate at velocity ranges of 2.0 to 3.0 fps.

* *Fish protection requirements* Some applications require velocities as low as 0.5 fps to minimize impingement of fish or other marine life.

* *Continuous operation* Screens that operate continuously may be able to justify higher velocities than those that operate intermittently.

The water velocity through the mesh is calculated as a function of the minimum water depth, maximum flow and the screen efficiency. Screen size selection is normally determined by solving for tray width according to the following general formula:

$$TW = Q / (LW)(POA)(V)(K)$$

Where: TW = Tray Width, ft
Q = Flow, gpm
LW = Water Depth, ft
V = Velocity through mesh, fps
K = 396 for Thru Flow Screens
= 740 for Dual Flow Screens
POA = % mesh open area (below)

Square Opening	W & M Gauge	Percent Open Area
3/16"	#18	0.6390
1/4"	#16	0.6380
1/4"	#14	0.5740
3/8"	#14	0.6790
3/8"	#12	0.6100

If a wire mesh with a rectangular opening, or a size is selected that is not listed above, the procedure for calculating the percent open area of the mesh is described in the Appendix.

It should be noted that this procedure for determining water velocity will yield the theoretical average velocity through the mesh. It does not consider velocity concentrations that may occur due to channel hydraulics, pump locations, or unique tray configurations. It is to be used as a guide
to assist in the determination of overall screen size. All velocity calculations should be verified by the screen manufacturer.

Consult manufacturer for all velocities greater than 2.5 fps, and tray widths larger than 10'-0".

Headloss

The differential headloss across a Traveling Water Screens is the difference in water level between the upstream and downstream sides of the ascending trays. The resulting headloss occurs as the flow of water in the channel is restricted by the wire mesh and screen structure.

The headloss is related to the water velocity through the mesh in general accordance with the formula:

$$\Delta H = (V^2 / 2g) SF$$

Where:
ΔH = Differential Headloss, ft
V = Velocity through mesh, fps
g = 32.2 ft/sec^2
SF = Safety factor

A Traveling Water Screens with a design velocity in the recommended range of 1 to 2.5 fps will have a clean screen headloss of less than 2".

The velocity through the mesh will increase as the wire mesh becomes fouled with debris. An increase in velocity results in an exponential increase in the headloss across the screen. It is therefore important that the design velocity of a Traveling Water Screens be considered at the extreme operating condition described above, and that it is low enough to be able effectively handle a sudden surge of debris. The presence of a headloss indicates that a severe or potentially severe condition may be developing. Any headloss in excess of 3" to 6" should result in operation of the screen.

The size and frequency of differential headloss occurrences has a direct bearing on screen operating costs. Traveling Water Screens operate under the same principal of operation that all other pieces of mechanical equipment do; the more frequent and severe the operation, the higher the operating cost.

As the headloss increases, the water pressure on the upstream trays increases at a rate of 62.43 lbs. per square foot of headloss area. This means that a 3.5 foot headloss occurring on a 10 foot wide Traveling Water Screens will result in a pressure of more than one ton (3.5 x 10 x 62.43) being applied to an otherwise balanced load. The increased pressure created by a headloss forces the tray chain against the chain track guides and increases the chain pull required to move the trays.

The increase in chain pull resulting from a differential headloss produces a proportional increase in the horsepower required to operate the Traveling Water Screens. The additional pressure caused by a headloss also accelerates wear of the screen chain, chain tracks, bearings, sprockets, and all other moving parts.

Potential headloss conditions should be considered when designing or specifying any Traveling Water Screens. Most Traveling Water Screens manufacturers design their "standard duty" screens so that they are able to start under the loads imposed by a 2'-6" headloss. "Heavy duty" applications, including those located on large rivers or in tidal areas, or those that may encounter heavy or unusual debris loads, should be designed to be able to start the screen under the loads imposed by a 5'-0" headloss. For applications considered "very heavy duty", it may be recommended that the screen be able to be started and operated continuously during a 5'-0" headloss.

A standard, 10'-0" wide Traveling Water Screens is usually capable of withstanding the loads resulting from a 5'-0" headloss. Unless special precautions are taken or specified, headlosses greater than 5'-0" may result in damage to the main framework, wire mesh, or tray frames.

Outdoor Operation

Traveling Water Screens are routinely installed outdoors with a minimum of weather-related problems. However, some precautions should be taken when any machinery is located where it is subject to severe weather conditions.

Common precautions, such as the use of proper lubricants during periods of temperature extremes, should be obvious. The mode of operation may also address potential weather-related problems. For instance, continuous, or frequent operation of Traveling Water Screens may prevent the buildup of ice on mechanical components.

Spray Wash System The primary cold weather concern with a Traveling Water Screen is the screen spray wash system. If sustained periods of very cold weather may be experienced, care should be taken to insure that the spray system is protected from freezing. The system should be equipped with provisions to automatically drain the screen spray pipe upon screen shutdown, and heat tracing of the spray water line is recommended.

Frazil Ice Intake screening equipment is susceptible to damage resulting from the formation and buildup of frazil ice.

The formation of granular ice crystals in turbulent, supercooled water is referred to as "frazil ice". Supercooled water occurs when the water temperature begins to drop and passes through the 32 degree Fahrenheit point. At a temperature of less than 32 degrees, sometimes as little as 0.1 degree less, tiny particles of ice form quickly and uniformly throughout the the water mass. Frazil ice may be extremely adhesive, sticking to any solid object that is at, or below freezing.

One method that has been used to solve the problems associated with the accumulation of frazil ice is the heating of the Trash Rack bars. Although it is necessary to heat the bars 0.2 degrees F above ambient, they are usually heated at least 2.0 degrees. Heating can be accomplished by resistive heating using heating elements imbedded in the bars, inductive heating using alternating currents to produce heat in strategically placed wiring, or by circulating heated fluids through passages in the objects to be heated.

Some electric power plants recirculate warm water from the discharge of their steam condenser to the mouth of the intake. This warm condenser exhaust discourages frazil ice formation and has been very popular. However it is becoming increasingly undesirable from the environmental standpoint.

Other methods that have been used with varying degrees of success include the use of plastic coatings to inhibit ice adherence, the use of non-ferrous materials (e.g. wood) with lower heat transfer properties, the use of log booms or breakwaters that calms the water and allows the formation of an insulating "ice cover".

SECTION 2

OTHER INTAKE SCREENS

DRUM SCREENS

A Drum Screen consists of a series of wire mesh panels mounted on the periphery of cylinder that slowly rotates on a horizontal axis. Drum Screens are reliable screening devices utilizing very few moving parts. The diameter of a Drum Screen may measure up to 50 feet or more, with effective screen widths of up to 14 feet.

Double Entry Drum Screen A Double Entry Drum Screen is mounted in a concrete channel with its axis perpendicular to the waters flow. Raw water enters the two open ends of the drum and flows through the wire mesh panels. Undesirable debris is retained on the inside surface of the wire mesh panels and is elevated out of the flow as the drum rotates. Debris retention is aided by lifting shelves that are fitted to the inside of the drum at regular intervals. The debris is removed from the mesh into a debris trough at the top of the drum by a high pressure spray wash system.

The drum structure consists of a central steel shaft fitted with a cast iron hub to which a series or radial arms, or spokes are attached. The central shaft rotates within sealed roller bearings mounted on joists that span the entrances to the drum. Structural steel cross members are fitted to the spokes to support the circular rims of the drum assembly. The circular rims form the framework to which the screening media is attached.

The drum is driven by means of a cast iron gear rack that is centrally mounted to the periphery of the drum. A driving pinion mounted at the operating floor elevation engages the gear rack and rotates the drum.

The rolled steel rim of the drum is designed to mate with a correspondingly shaped angle bolted to the concrete intake walls to seal the drum. The seal may be fitted with a neoprene seal to provide sliding contact, and a tighter seal.

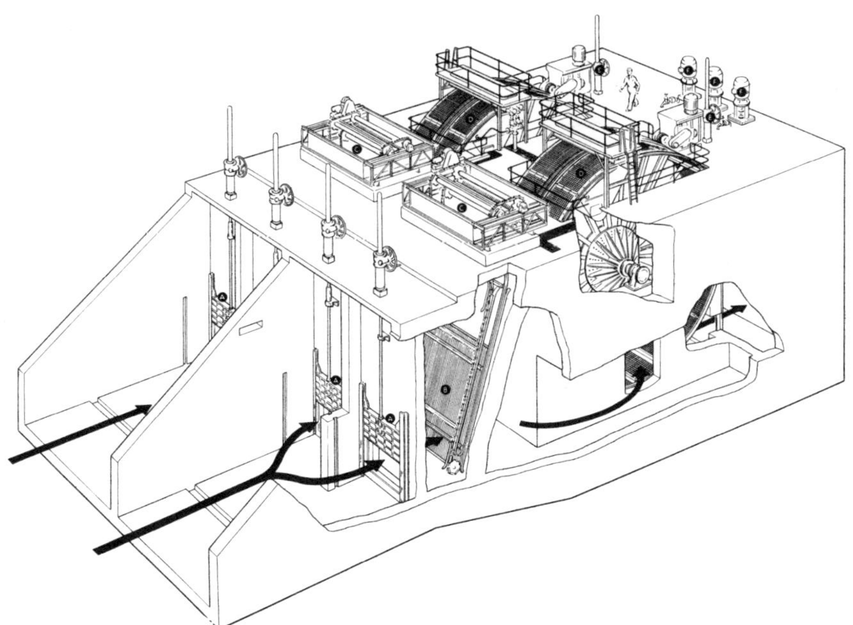

Drum Screen Installation
Courtesy of William Green, Ltd.

The civil costs and the initial price of a Drum Screen installation is higher than that of a Traveling Water Screen. Because of the simplicity of design however, the maintenance and operating costs of a Drum Screen are usually less those of a Traveling Water Screen.

Drum Screens are used widely throughout Europe and South America in electric power plants and other large water intakes, but in the United States their use is primarily limited to irrigation projects.

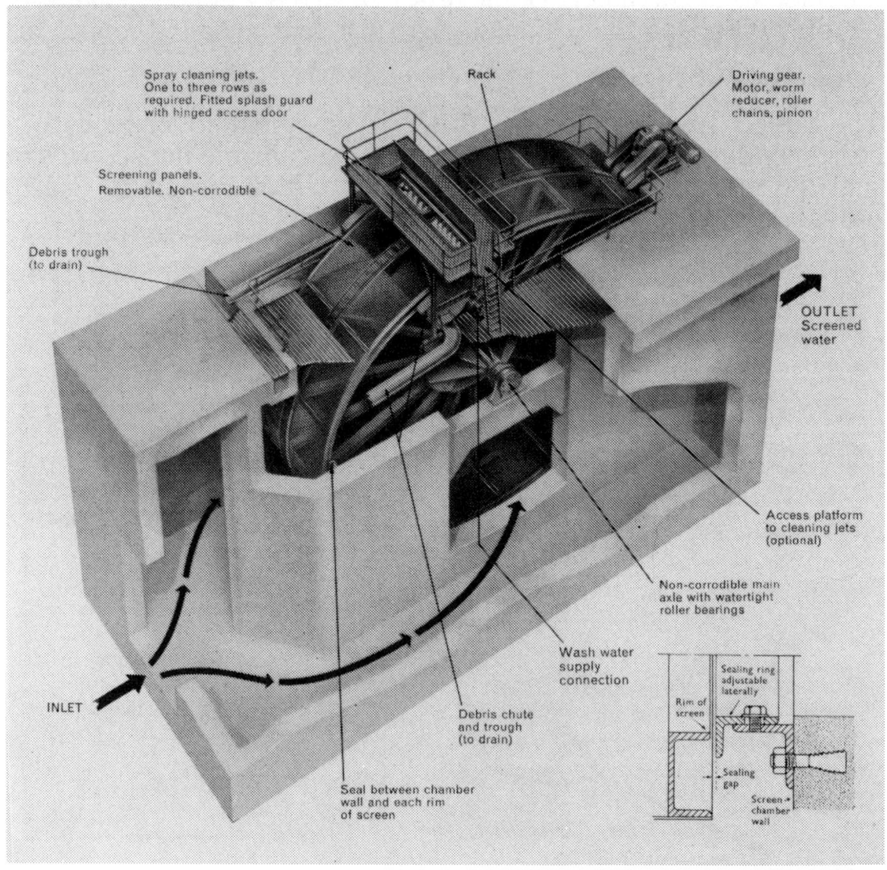

Typical Drum Screen Installation
Courtesy of William Green, Ltd.

Drum Frame Construction
Courtesy of Hubert Water Systems BV

Single Entry Drum, or Cup Screen Single Entry Drum Screens, also called Cup Screens, are similar in operation to the larger Double Entry Drum Screens. The major difference lies in their orientation to the flow and the fact that the downstream end of the drum is closed with a full-diameter backplate. The central shaft of the Cup Screen is located parallel to the flow and the open end of the drum faces the flow.

Flow enters the drum through its open end and exits outward through the wire mesh mounted on the screens periphery.

Cup-Type Drum Screen Installation
Courtesy of McGinnes/Royce, Inc.

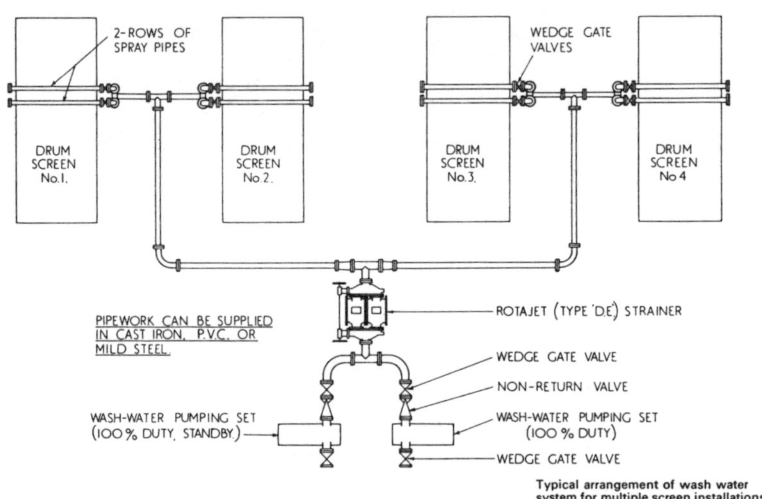

Multiple Drum Spray Wash System
Courtesy of William Green, Ltd.

The gear rack used to drive the screen is mounted to the backplate, and an adjustable thrust bearing is usually located on the downstream side of the screens central shaft to absorb thrust loads.

Cup Screens are available in 5 to 25 foot diameters and widths of 1 to 8 feet. For flows of 20,000 gpm or less, tank mounted, package units are available.

Cup-Type Drum Screen Frame Construction
Courtesy of FMC Corporation

REVOLVING DISC SCREENS

The Revolving Disc Screen is a simple and compact screening device. It consists of a flat disc covered with screening media that rotates about a horizontal axis, perpendicular to the water flow. As water flows through the submerged portion of the disc, solids are retained by the screening media. The rotation of the disc lifts the solids above the water surface where they are removed by a bank of spray nozzles.

Disc Screens are available in sizes ranging from 6 to 16 feet in diameter with mesh openings as fine as 1/32". Applications are limited to those with flows less than 20,000 gpm, having small water level fluctuations.

Disc Screen
Courtesy of FMC Corporation

Disc Screen Frame Construction
Courtesy of FMC Corporation

PASSIVE SCREENS

Passive Screens are stationary screen cylinders that rely on ambient currents and velocity control to minimize debris buildup. The periphery of the cylinder provides the screening surface consisting of a trapezoidal shaped "wedge wire" bar formed to maintain a uniform 0.040 to 0.50 inch screen opening. The cylinders may range in size from 12 to 84 inches in diameter, and 14 to 275 inches long.

Passive Screen cylinders are normally mounted on a horizontal axis, oriented parallel to the ambient currents which provide necessary cleaning action. Screens may be equipped with a 100 to 150 psig airburst backwash system to aid in the cleaning of the screen surface.

Most installations are designed for a maximum intake velocity of 0.5 feet per second through the screen opening.

There are many intake configurations available to suit a variety of application requirements. Small installations may use a single screen, a single tee unit with two screens, or a multi-tee system. Large installations may utilize multiple screen arrays manifolded to provide equal flow through each screen.

Typical Passive Screen Arrangement
Courtesy of Johnson Screens

A tower arrangement may be provided with screens mounted at various elevations. This will allow the operator to select the level at which water will be drawn to accomodate seasonal changes in water level, quality or temperature.

Periodic physical cleaning may be required due to debris accumulations or biofouling, especially on marine intakes. If frequent cleaning is anticipated, bulkhead mounting of the screens is recommended to provide easier access. Bulkhead mounted units are located over inlet ports on vertical rails to facilitate their removal.

Marine intakes can be fabricated of copper-nickel material to resists fouling through a slow, steady release of copper that is toxic to potential encrustants.

Passive Screen Tower Arrangement
Courtesy of Johnson Screens

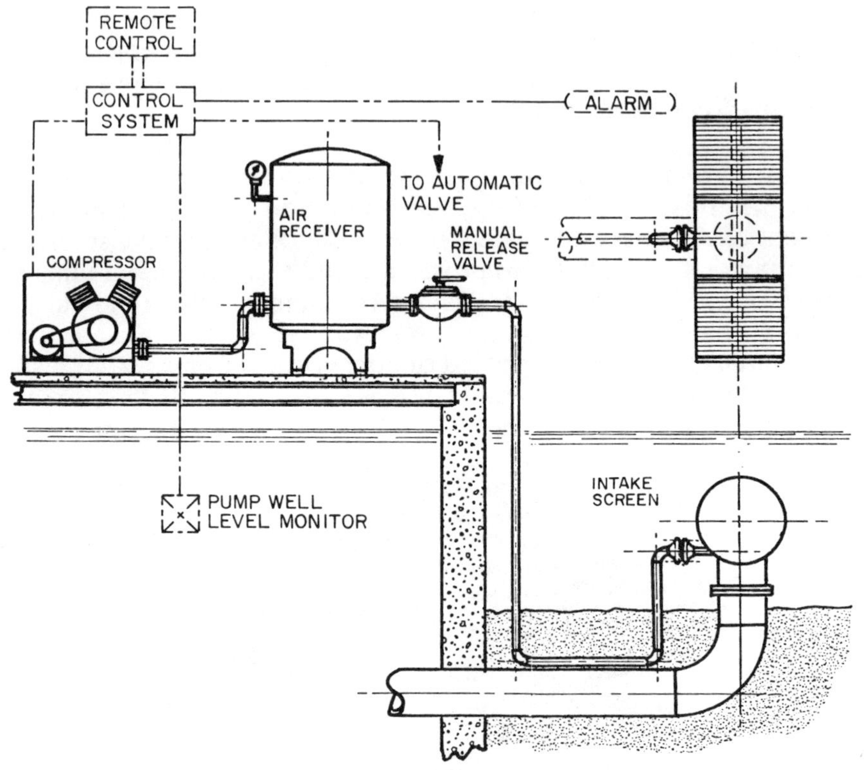

Passive Screen Backwash System
Courtesy of Johnson Screens

RADIAL WELL SCREENS

A Radial Well Intake is an infiltration system that utilizes slotted pipes placed horizontally in a naturally occurring sand or gravel aquifer. Most installations utilize several pipe screens that radiate outward from a vertical pump sump. Water enters the system through the screen pipes and empties into the central sump where it is removed by pumping.

The intake is constructed by installing a vertical pump sump below the water table near a river or creek with radial wells jacked out into the surrounding aquifer. They are "developed" in the same manner as conventional wells.

The use of this intake system is limited to locations that have a permeable substrate and reasonably clean water. The maximum capacity of a single well is approximately 20,000 gpm, although several wells may be placed in a group to provide increased flow.

Where a combination of appropriate geology and flow capacities exist, this intake competes favorably with conventional intake designs. This design also provides reduced operating expenses than conventional wells due to lower pumping heads and greater pumping efficiencies.

SECTION 3

FISH SCREENS

Section 316(b) of the Federal Water Pollution Control Act of 1972 specifies that the cooling water intake structures for all industrial categories:

> "require that the location, design, construction, and capacity of cooling water intake structure reflect the best technology for minimizing adverse environmental impact."

The environmental impact most often associated with cooling water intakes are the effects of impingement and entrainment of marine life within the intake system.

The entrapment of fish and other marine life against the screening media is referred to as "impingement". Impingement usually results when organisms cannot escape the area in front of the screen because of the velocity of the intake stream. Impingement is usually fatal, as it leads to exhaustion and/or physical damage of the marine life.

Entrainment is the incorporation of small organisms, including the eggs and larvae of fish and shellfish, into the intake system. Entrained organisms pass through the screen media with damage resulting from mechanical, temperature or pressure stresses associated with other downstream equipment or systems.

There are several design considerations and modifications of conventional Traveling Water Screening equipment that can significantly reduce adverse environmental effects of a raw water intake system. These modifications are known collectively as "Fish Screens". Some Fish Screen features and technology are proprietary, and most are applied on a site specific basis.

Every water intake is unique. Impingement and entrainment problems may be related to particular season of the year, fish species, tidal condition, intake location, or water velocity.

It is generally accepted that the most desirable features of an intake utilizing Traveling Water Screens are those that provide:

* *low approach and throughflow velocities (<1 fps)*
* *open or short intake channels with "escape routes"*
* *small mesh openings*
* *provisions to gently handle impinged fish*
* *continuous operation*

Each of these features add to the cost of the intake construction and operation. It is important to determine the feature or combination of features that result in the most cost effective Fish Screening system.

CONTINUOUS OPERATION

Most conventional screen installations are designed for intermittent operation. However, an essential feature of any fish protection system is its ability to operate continuously.

It has been demonstrated that continuous operation of Traveling Water Screens alone, without any special fish handling provisions, will reduce impingement and entrainment of marine life. This can be attributed to the fact that continuous operation will prevent accumulations of debris that result in increased velocities and a subsequent differential headloss. As headlosses and velocities increase, it is far more likely that fish will not be able to escape the screen area, become impinged, and die.

If a screen has been designed with special fish handling provisions, it is important that the screen be operated continuously to remove impinged fish as quickly as possible. The shorter that the organisms are subjected to the stresses of impingement, the greater their chances of survival.

If a Traveling Water Screen is to be operated continuously, careful attention should be given to the selection of the mechanical components. The following items warrant special consideration:

* *Footshaft bearing or bushing material.* Standard bronze or Ryertex bushings may be replaced with a special centrifugally cast, wear resistant alloy bushing.

* *Footshaft assembly design.* A split shaft design will allow easier replacement or repair of footshaft assembly components.

* *Headshaft takeup bearing.* Some manufacturers recommend the use of anti-friction takeup bearings as replacements for standard.bronze bushed bearings.

* *Chain tensioning system.* A chain tension indicator or an automatic chain tensioning device will aid in maintaining proper chain tension.

* *Drive unit.* The drive components must have a service factor that reflects continuous operation.

* *Tray travel speeds.* Two speed operation will reduce wear. A slow speed can be used for normal operation, and a higher speed can be used during periods of high debris or fish loading.

* *No-motion switch.* This switch advises the operator if screen trays are not revolving.

* *Differential Controller.* This controller will only operate screens at "high speed" when debris loading indicated that it is necessary.

* *Chain joint material.* The material selection of the chain pins, rollers and bushings should provide corrosion and wear resistance.

Continuous screen operation will also require that a firm inspection and maintenance schedule be prepared and followed. In addition to the above components, a routine inspection of the tray attachment bolts, headsprocket tooth inserts, nozzle orifices, and drive unit is required.

INTAKE LOCATION

The location of an intake is the first consideration necessary when designing a system to minimize fish impingement or entrainment. Most existing designs have been based on engineering and economic factors with little regard to the protection of fish and other aquatic biota.

Biological field studies may provide important clues in determining the most practical method of minimizing the adverse effect of intake location on the aquatic life in the water source. The type and location of the intake should be considered relative to the water level and topography of the site. Other factors that must be reviewed include the location of plant discharge, navigational routes, recreational areas, ease of construction, and aesthetics.

Improper location of an intake may actually attract marine life. Many "conventional" designs hopelessly trap fish in long dead end channels. A seemingly minor relocation of an intake to another location on the same plant site and water body may avoid areas of high fish concentrations.

FRONT DISCHARGE FISH SCREENS

There are two popular and reasonably economical alternatives to conventional Traveling Water Screens. The front discharge Fish Screen is one of two alternatives that have been used successfully in the United States.

Front Cleaned Fish Screen
Courtesy of FMC Corporation

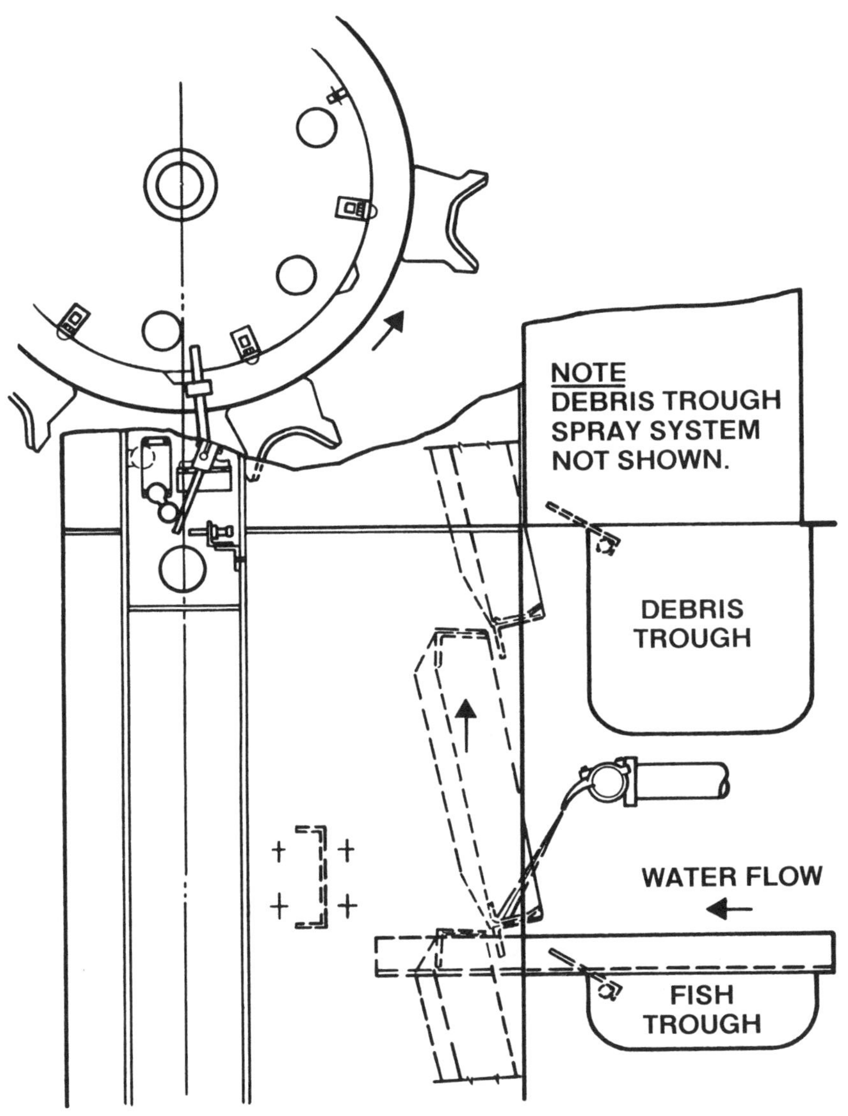

Front Cleaned Fish Screen Schematic
Courtesy of FMC Corporation

This front discharge Fish Screen is equipped with 1-3/4" deep, watertight fish pans mounted on the lower trash shelf of each tray. As the trays revolve and are lifted out of the water, fish and other marine life are retained in the fish pans. The pans have been designed so that a low pressure spray system mounted outside the tray line displaces the water in the pan and flushes its contents into a fish sluice trough located on the front, upstream side of the screen.

After the pans have been emptied, a conventional high pressure spray washes debris from the wire mesh into a separate debris trough, also located on the front, upstream side of the screen.

Many existing Traveling Water Screens can be retrofit to provide these fish removal features.

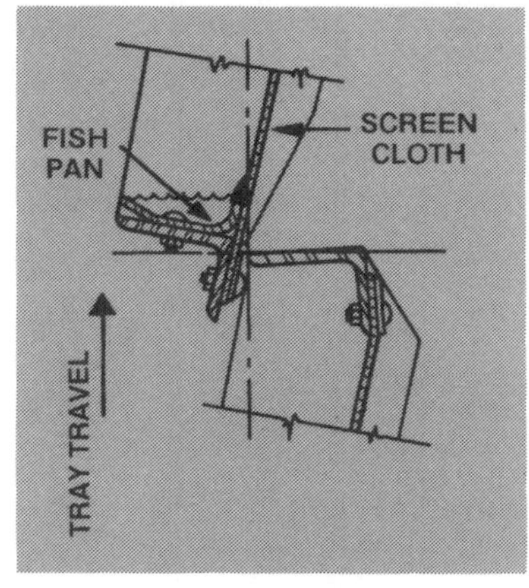

Front Cleaned Nozzle and Fish Tray
Courtesy of FMC Corporation

REAR DISCHARGE FISH SCREENS

The rear discharge Fish Screen is also equipped with watertight fish buckets mounted on the lower trash shelf of each tray. These buckets collect and retain marine life in 2 to 4 inches of water until the trays revolve over the headsprockets and spill their contents, with the aid of gentle sluice sprays, into a fish trough for later release. The trays continue past a high pressure spray system to remove any remaining debris into the debris trough for further disposal.

When considering a rear discharge Fish Screen, special attention should be given to the tray and fish bucket design. It is extremely important that the tray design provides a barrier-free discharge of marine life into the fish trough. In most conventional tray designs, the screening media is mounted at an angle of approximately 8 degrees in the direction of the flow. This incline, coupled with some structural tray frame designs, produces a ledge or pocket that may trap or hinder the discharge of marine life as the trays revolve around the headsprockets.

Inclination of the entire screen frame will also facilitate rear discharge. If any obstruction prevents a smooth rear discharge of marine life, the trays must be redesigned or modified with individual deflector plates.

Some fish bucket designs may actually increase the damage of impinged fish. While submerged, water flowing over the leading edge of the fish bucket may cause a severe vortex that may repeatedly slam fish into the wire mesh causing them irreparable damage.

Many conventional Traveling Water Screens designs can be modified to include a rear discharge fish recovery system. As a minimum, the modifications will include the addition of fish buckets to each tray assembly, the addition of a low pressure fish spray, the addition of a fish sluice trough, and modifications to the housings. If the existing screen utilizes a front discharge spray system with a below grade debris trough, it may be necessary to increase the height of the screen headshaft to accommodate both troughs on the rear of the screen.

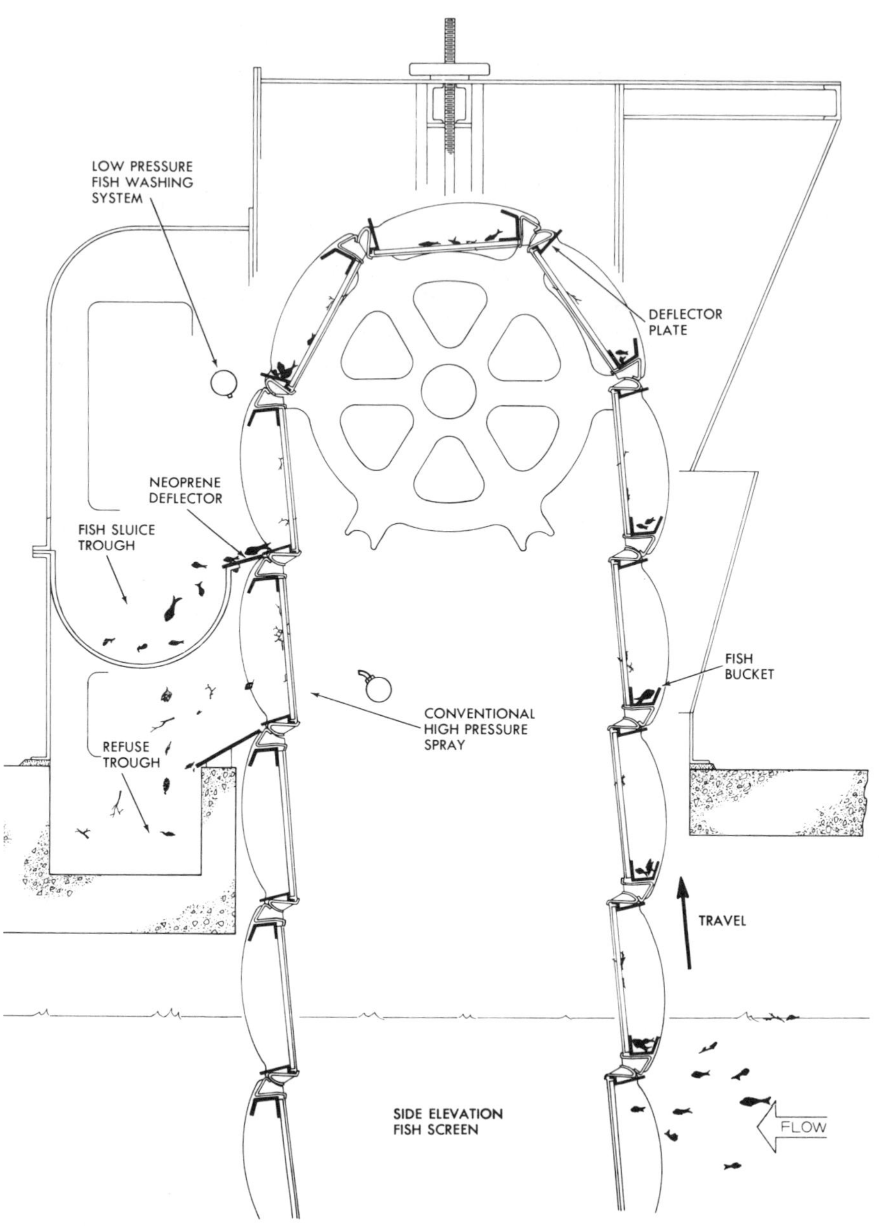

Rear Discharge Fish Screen
Courtesy of Envirex, Inc.

Modifications of existing screens are highly site specific and should be carefully considered with the manufacture. It may be more cost effective to completely replace the existing screens than it would be to modify them.

FLUSH MOUNTED FISH SCREENS

Many conventional Thru Flow and Dual Flow Traveling Water Screens have been mounted near the downstream end of an intake channel. Fish, or other marine life entering the channel may become trapped in the confined channel due to the increased channel velocities. They eventually being impinged on the screen mesh as they tire and are no longer able to fight the current.

This problem may be overcome by mounting a Thru Flow screen so that the face of the screening surface is virtually flush with the shore line. Fish nearing the screen may be able to escape much easier by moving to the left or right of the screen face.

Coarse trash racks can be located out in front of, and on the sides of the flush mounted screens to prevent damage from large debris, while preventing large enough openings for the fish to swim through freely.

PLATFORM MOUNTED FISH SCREENS

Another design that eliminates the use of confining channels has a Dual Flow (double-entry/single-exit) screen mounted on a platform or pier. The screen simply "hangs" down into the water from the platform. Water passes through the screen baskets and into the center of the screen. It exits through a port in the screen framework and into a plenum that may be connected directly to the pump suction.

This screen may also be surrounded by a coarse bar grid to prevent damage from large debris.

RADIAL WELL INTAKES

A Radial Well intake is an infiltration system that utilizes slotted pipes placed horizontally in a naturally occurring sand or gravel aquifer. Most installations utilize several pipe screens that radiate outward from a vertical pump sump. Water enters the system through the screen pipes and empties into the central sump where it is removed by pumping.

Radial Well intake systems have the environmental advantage of segregating the marine life populations from direct contact with the screen mechanism or screen structure.

PASSIVE SCREENS

Passive Screens are designed to prevent debris buildup through the use of flow and velocity control. These features also assist in the reduction of impingement and entrainment of marine life.

The stress experienced by marine life during the mechanical handling involved with other screening processes is eliminated, and the finer openings and low velocities associated with passive screening result in reduced impingement and entrainment.

HORIZONTAL SCREENS

Horizontal Traveling Water Screens have been developed for use in applications with water levels less than ten feet deep. These screens consist of a continuous series of vertical trays fitted with wire mesh that travel in an elliptical path. This screen resembles a trolley conveyor with trays that are carried by an endless power driven strand of chain suspended from an overhead track.

Horizontal Screens have been designed primarily for use in conjunction with fish bypass systems, and have been used with limited success. They are not currently recommended for general debris removal applications.

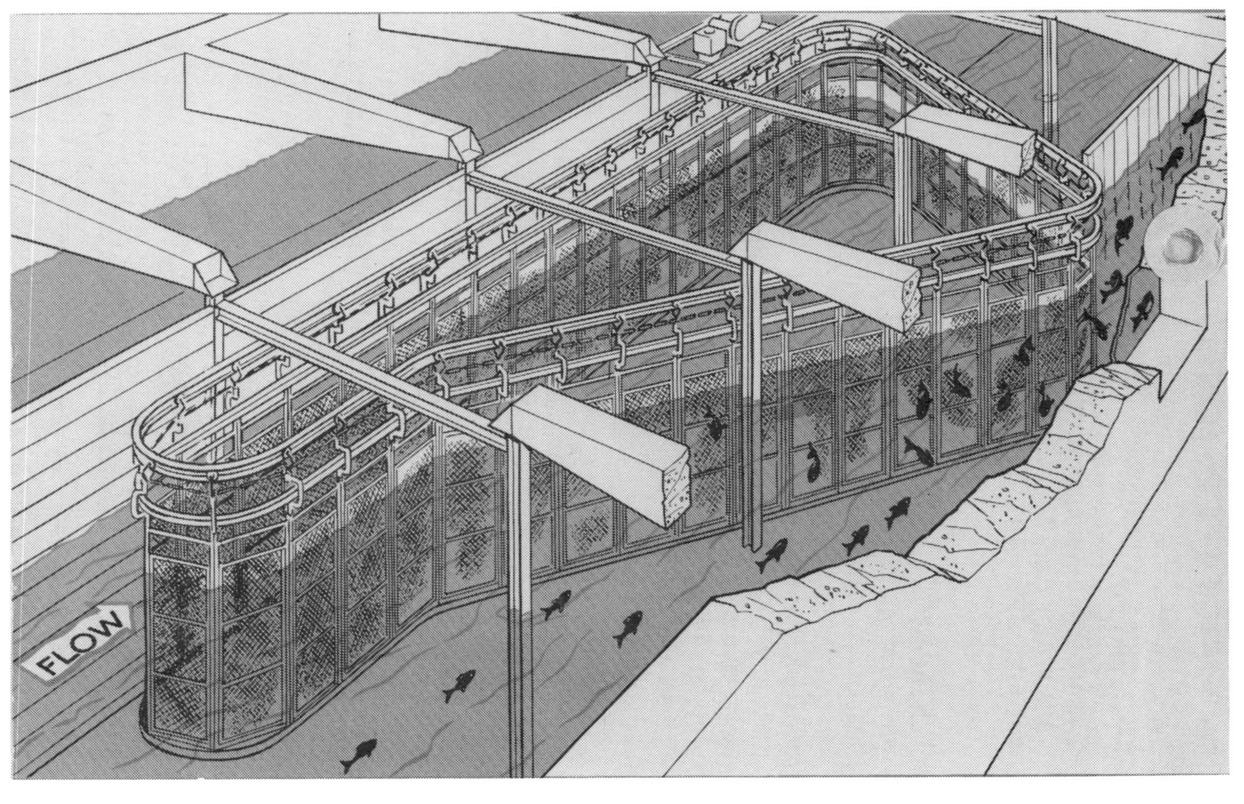

Horizontal Screen
Courtesy of Envirex, Inc.

ANGLED FISH DIVERSION SCREEN

Thru Flow Traveling Water Screens may be mounted at an angle to the flow to assist fish to an escape area by a velocity component of the flow. This arrangement may create an area of turbulence along the screen face that fish will avoid as they are directed to the escape bypass.

As they reach the bypass, fish may be transported back to the water body by jet pumps, non-clog centrifugal pumps, airlift pumps, screw pumps, or a mechanical lift or "elevator" bucket.

An Angled Fish Diversion system requires an extensive civil structure and may be cost prohibitive for most applications. It has also been demonstrated that this concept looses effectiveness as the size of the intake increases.

BEHAVIORAL BARRIERS

There are a number of behavioral barriers that have been used to reduce impingement and entrainment of marine life at water intakes. These include electrical, sound, light, water jet, air bubble, stationary chain, and louver barriers. These methods have met with varying degrees of success, usually limited to specific species and sizes of fish, and particular seasons of the year.

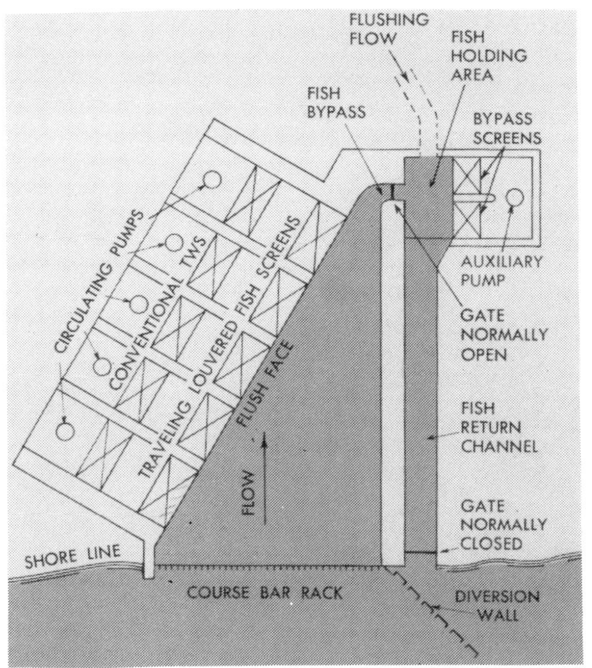

Angled Fish Diversion Screen Arrangement
Courtesy of Envirex, Inc.

SECTION 4

BAR SCREENS

A Bar Screen consists of a stationary bar rack that is automatically cleaned by one or more power operated rakes. As a rake is operated up the face of the bar rack, it removes accumulated debris and elevates it out of the flow. At the top of its operating cycle, debris is swept from the rake into a trash receptacle by a pivoting wiper mechanism.

Bar Screens are most frequently used at the headworks of wastewater treatment plants to remove material that may interfere with, or not respond to accepted methods of treatment. Because they are the first unit process in the treatment plant, they play an important role in determining overall plant performance. Bar Screens are also used to remove large solid objects, rags, and other debris that may clog piping and damage pumps or other downstream equipment at pump stations, flood control projects, and irrigation facilities.

Industrial plants often employ Bar Screens to reduce the suspended solids levels before discharging plant effluent into a municipal sewer. In some industrial installations the screens may be used to recover sufficient materials to pay for their installation and maintenance.

There are design variations available to suit most screening requirements and operator preferences. Manufacturers offer standard screens with bar spacings that range from less than 1/4" to 4" for use in channels up to 14 feet wide and more than 80 feet deep. Bar Screens can be divided into three categories: "Trash Rakes" with bar spacings that range from 1" to 4", "Mechanically Cleaned Bar Screens" with bar spacings that range from 1/2" to 1-1/2", and "Fine Screens" with bar spacings less than 1/2".

TRASH RAKES

Trash Rakes are heavy duty raking devices that are used to remove rough or heavy debris retained on a trash rack having 1-1/2" to 4" bar spacings. Trash Rakes are frequently used as a primary screen ahead of the finer mesh Traveling Water Screens at electric power plants, or to protect pumping equipment at pump stations and flood control projects.

Their are several Trash Rake screen designs with effective rake widths of up to 20 feet. Trash Rakes may be furnished as stationary units dedicated to clean a single trash rack, or wheel-mounted to traverse the entire width of an intake structure and clean individual sections of a very wide trash rack.

Because of the large variations in the bar spacings, debris loading, and application of Trash Rakes, there is no established relationship for the quantity of screenings removed per unit volume of flow.

CABLE OPERATED TRASH RAKES

This Trash Rake design utilizes a single, toothed rake that is raised and lowered by means of one or more sets of stainless steel cables, to clean a stationary trash rack. The rake is lowered down the

Cable Operated Trash Rake
Courtesy of Jeffrey Division, Dresser Industries

upstream face of the trash rack with the teeth in the open, disengaged position. As the rake reaches the channel bottom, the teeth are rotated to their closed position, meshing with the stationary bars. The rake is then lifted up the face of the trash rack, collecting accumulated debris.

As the rake reaches the discharge position above the operating deck elevation, the teeth are opened, discharging debris into a trough or hopper.

Rake Hoist The upper ends of the 3/8" minimum diameter rake cables are fixed to a shaft-mounted hoist drum. The drum's line shaft operates in self-aligning roller bearings and is directly coupled to a drive unit furnished with a motor mounted brake. Rotation of the hoist drum changes the effective length of the cables and results in the raising or lowering of the rake. The excess cable that accumulates during the rakes ascent is stored on the periphery of the grooved hoist drum. Limit switches automatically stop the rake at the top and bottom of its travel.

Hoist speeds may range from 15 to 35 feet per minute. Two speed hoist drives (slow descent, fast ascent) may be considered for deep trash rack installations to reduce rake cycle times.

Standard duty Trash Rakes are furnished with a single hoist drum and one set of cables to lift objects up to 17" in diameter with a hoist capacities to 2,000 pounds. Heavy duty units may be capable of lifting objects up to 24" in diameter, and are furnished with dual hoist drums, two sets of cables, with hoist capacities to 4,600 pounds.

The hoist mechanism should be equipped with a slack cable safety switch and an over-travel switch. The hoist drive motor of Trash Rakes typically ranges from 1 to 7.5 horsepower.

Rake Tooth Operation The rakes cleaning cycle begins when the rake is lowered to the bottom of the channel and the rake teeth are rotated to the closed, engaged position. It is recommended that the rake teeth penetrate the bar rack a minimum of 1/2".

Tooth opening and closing is also accomplished by changing the

effective cable length. This may be done by controlling the slippage between a fixed and free segment of the hoist drums. This system requires that the inside drum, or drum segment be keyed to the drum shaft. The other drum is free for limited rotation controlled by the fixed drum. The free drums are equipped with band brakes to allow for the opening of the rake.

An alternative arrangement for operating the rake teeth utilizes a walking beam assembly. This system requires a small hydraulic power unit to operate a hydraulic cylinder that is used to raise and lower a walking beam assembly, changing the effective length of the tooth control cables. Hydraulic relief valves automatically open the rake teeth if an overload condition occurs.

Guided Rake Each end of the rake assembly may be equipped with wheels or sliding blocks to operate in guide channels embedded in the walls of the concrete channel.

The guided rake design is recommended for vertical trash racks or trash racks inclined less than 10 degrees from vertical.

Non-Guided Rake The non-guided rake design allows the rake to ride over obstructions on the trash rack, and is preferred for most Trash Rake installations. This design requires the rake assembly to be equipped with flanged wheels, each wide enough to ride on two bars.

Non-guided rake designs are recommended to be used with trash rack inclinations of 10 degrees or more.

Traversing Mechanism Trash Rakes can be mounted on a self-propelled hoist frame capable of traversing the width of an entire intake. This allows the Trash Rake to be positioned to clean a specific portion of a wide trash rack.

A 1.5 to 3 horsepower drive unit can propel the screen hoist frame at a speed of approximately 25 feet per minute on flanged wheels riding on 40 lb./yd. tee rails with a rail gauge of 3'-6" to 4'-6". Tie-down clamping brackets are used to engage the downstream rail and prevent the unit from tipping as large debris loads are lifted. The hoist frame is usually equipped with an operators platform and pushbutton station from which the unit is operated. A pendant control station may be furnished.

The rakes transition from the trash rack to its discharge position on the hoist frame is accomplished through the use of a deadplate constructed as an integral part of the hoist framework. The deadplate should be designed to operate over a 3'-6" high handrail.

A spring loaded electric cable reel with sufficient cable to span the intake structure must be provided to store the electric power cable.

Traversing Trash Rake Mechanism
Courtesy of Envirex, Inc.

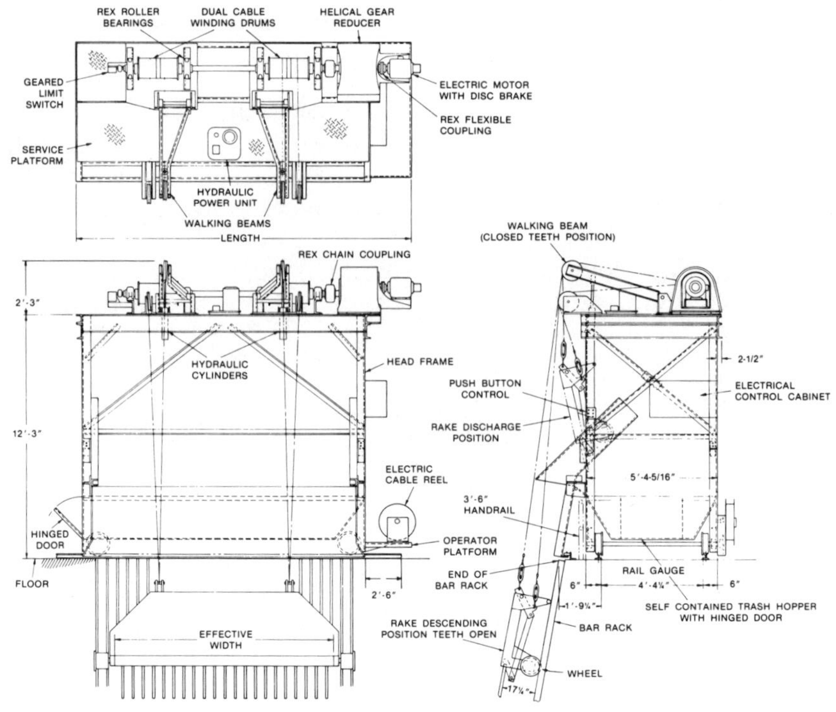

Traversing Trash Rake
Courtesy of Envirex, Inc.

Debris Trough or Hopper Debris collected by the rake can be discharged into a trough or debris hopper located on the downstream side of the screen.

Debris hoppers are commonly fabricated as an integral part of the hoist frame. This type of hopper should be furnished with a hinged bottom or a side door for debris cleanout.

The debris hopper may be wheel or rail mounted, and designed so that it can be removed from the frame entirely. Commercial "dumpsters" may also be utilized as Trash Rake hoppers.

Traversing Trash Rake Mechanism
Courtesy of FMC Corporation

HYDRAULIC TRASH RAKE

A hydraulically operated rake mechanism may be used to clean trash accumulated on the face of a trash rack. There are several variations of Hydraulic Trash Rakes available with lifting capacities of up to 2500 pounds, for use with trash racks up to 65 feet deep.

These units consist of a single rake that can be raised and lowered and opened and closed by means of two or more hydraulic cylinders. Hydraulic Trash Rakes can be furnished as stationary units or can be motor driven to traverse the width of an intake.

Hinged-Arm Type The operation of this Trash Rake resembles that of a hydraulically operated excavator or backhoe. The synchronized motion of the hydraulic cylinders is used to extend the rake out in front of the trash rack, lower it to the channel bottom, and lift debris out of the water for disposal at the deck elevation. The operator may control the movement of the pivoting-arm rake mechanism with a pendant-type hand controller or stationary push button station, or while sitting in a fully-enclosed cab equipped with a joystick-type controller.

Hydraulically Operated Trash Rake
Courtesy of Smalley Excavators, Inc.

Hydraulically Operated Trash Rake
Courtesy of Acme Engineering

Some designs allow the upper unit to rotate 360 degrees so that debris can be lifted out of the channel and unloaded in any direction.

Although they may be stationary, most of these rake units are mounted on tee-rails to be electric driven along the width of the intake. They may also be furnished with rubber tires and a diesel or gasoline powered engine that allows them to be driven to other locations. Other available options include climate controlled cabs, and various grab or clamshell rake bucket configurations.

Maximum rake widths vary with specific screen designs ranging from 4 to 12 feet. Electric drive motor sizes range from 5 to 30 horsepower. Diesel operated units may range to 45 horsepower.

Telescoping Arm Type Another Trash Rake design consists of a single rake attached to the bottom of a carriage-mounted telescoping arm. The arm may be raised and lowered by means of a chain lift mechanism or a hydraulic cylinder. Engagement and disengagement of the rake is accomplished by a hydraulic cylinder that tilts the rake carriage into, or away from the trash rack.

The carriage may be mounted on a horizontal rail to traverse the width of the intake.

Telescoping Arm Type Trash Rake
Courtesy of Acme Engineering Company

Hydraulically Operated Trash Rake
Courtesy of Cross Machine, Inc.

Hydraulically Operated Trash Rake
Courtesy of Atlas Polar, Hercules Hydrorake Division

CHAIN OPERATED TRASH RAKE

Chain Operated Trash Rakes utilize toothed rakes or debris lifters mounted on endless strands of chain and operating over head and foot sprockets. The rakes clean the bar rack on their ascent discharge debris on the top, downstream side of the screen prior to beginning their descent.

These rakes are available in front and back cleaned designs.

Chain Operated Trash Rake
Courtesy of Ossberger Turbines, Inc.

Chain Operated Trash Rake
Courtesy of Ossberger Turbines, Inc.

Chain Operated Trash Rake
Courtesy of Duperon Corporation

MECHANICALLY CLEANED BAR SCREENS

Mechanically Cleaned Bar Screens are automatically operated screening devices that are used primarily to screen raw sewage at the headworks of a waste treatment plants. Although some plants still rely on manually cleaned bar screens, a Mechanically Cleaned Bar Screen will increase plant efficiency by removing debris accumulations as they occur. This reduces the possibility of sewage backups and grit sedimentation that occur if a screen is plugged, and the subsequent influent surge that accompanies cleaning a plugged screen.

There are many variations of Mechanically Cleaned Bar Screens that are designed to suit virtually any screening requirement with bar spacings that usually range from from 1/2" to 1-1/2". The operation of a Bar Screen can be controlled to insure that the bar racks are cleaned intermittently or continuously, and that the screenings are contained in an appropriate receptacle.

The quantity of screenings removed is a function of the width of the openings in the bar rack, although typical Mechanically Cleaned Bar Screen installations yield 0.5 to 10 cubic feet of debris per million gallons of screened sewage.

Multi-Raked Bar Screens

Multi-Rake Bar Screens utilize multiple, chain-mounted rakes to clean a single, stationary bar rack.

The primary advantage of Multi-Rake Bar Screens is its ability to quickly and continually remove large volumes of debris from the face of the bar rack. The time interval that a rake passes over a given portion of the bar rack is reduced through the use of multiple rakes. Based on a typical rake speed of 10 feet per minute, and a typical rake spacing of 8 feet, a "new" rake will pass over a given portion on the bar rack every 48 seconds. This short "cycle time" is important with deep channels or those applications that may experience periods of heavy debris loading.

The major disadvantage of the multi-rake designs is the quantity and location of many of the mechanical components. These designs typically use a series of rakes that are mounted on two endless strands of chain, operating over head and foot sprockets A portion of the chain (and usually other mechanical components including sprockets, bushings, shafting) is operated underwater. The makes inspection and maintenance of the equipment, especially the submerged components, difficult.

Power is usually provided by an electric motor driving a helical or worm gear type speed reducer. Screens are available with shaft mounted drive units directly mounted on the screen headshaft, or with a roller chain drive operating over a drive/driven sprocket arrangement. Rake travel speeds range between 7 and 15 feet per minute.

FRONT CLEANED, FRONT RETURN BAR SCREENS

A front cleaned, front return Bar Screen consists of a series of chain-mounted rakes that revolve around head and foot sprockets, to clean an inclined bar rack. This model Bar Screen has been used extensively on "standard" duty sanitary sewage flows.

Both the ascending (cleaning) and descending (return) runs of chain are located on the front, upstream side of the bar rack. Front cleaning provides positive cleaning of the front and sides of the bars by the toothed rakes, and front chain return eliminates the possibility of debris carryover to the downstream side of the screen.

The screen chain is generally 6" pitch pintle type (e.g. #720 or 720S) operating over cast head and foot sprockets with hardened teeth. Split-type headsprockets are recommended to provide ease of assembly and disassembly.

The screen headframe is fabricated from 3/16" minimum thickness steel, and the sideframes below the floor level are fabricated of steel having a minimum thickness of 5/16". The sideframes are formed to provide u-shaped guides for the ascending and descending chains, and include shrouds around the footsprockets to prevent debris from becoming caught between the chain and sprocket teeth. A curved bootplate, recessed in the channel bottom, provides a bottom seal and a smooth transition as the descending rakes revolve around the footsprockets and begin their ascent.

Front Cleaned-Front Return Multiple Rake Bar Screen
Courtesy of Envirex, Inc.

Debris is removed on the downstream side of the screen by a pivoting rake wiper mechanism. The mechanism should be equipped with adjustable shock absorbers to regulate the return of the wiper.

The front cleaned, front return Bar Screen is usually installed in a concrete channel at an inclination of 6 to 30 degrees from vertical in the direction of the flow. The channel may include guide slots cast into the concrete walls that are arranged to receive the screen framework. The recessed frame will allow for the bar rack's complete use of the channel width.

An alternative method of installation is to mount the screens headframe on pillow block bearings located above the operating floor. The entire screen may then be pivoted out of the channel for maintenance or inspection.

Front Cleaned Bar Screen Pivoted for Maintenance
Courtesy of Envirex, Inc.

Combination Bar/Grit Screen At least one manufacturer has adopted a front cleaned, front return Bar Screen to remove both screenings and grit in a single unit. This screen can eliminate the need for separate screen and grit removal systems in small and medium sized treatment plants having normal capacities of up to 2.75 mgd.

This screen utilizes an adjustable baffle and a steeply inclined grit sump to promote settling of grit at the base of the screen. The cleaning rakes are fitted with perforated buckets that collect grit from the screen sump as they revolve around the footshaft. A "knocker" is used to aid in removal of grit from the bucket, and a pivoting wiper mechanism cleans the teeth of any material that does not drop off.

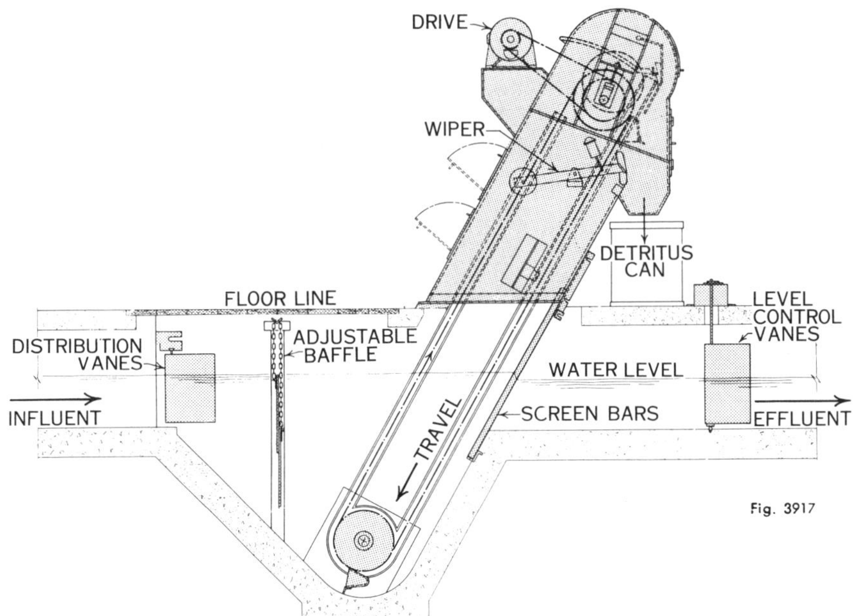

Screen width, feet and inches	Capacity M.G.D.		Water depth thru screen in feet and inches			
			Screen bar width, inches			
	Normal	Storm	¼		⅜	
			Clear openings between bars, inches			
			¾	1	¾	1
2-0	.50	1.00	0—11	0—11	0—11½	0—11½
2-6	1.10	2.20	1—1½	1— 1	1— 2	1— 2
3-0	1.50	3.00	1—2½	1— 2	1— 3	1— 2½
3-6	2.00	4.00	1—3	1— 3	1— 4	1— 3½
4-0	2.75	5.50	1—4½	1— 4	1— 6	1— 5

Diagram of Combination Grit/Bar Screen
Courtesy of FMC Corporation

Combination Grit/Bar Screen
Courtesy of FMC Corporation

This combination grit/bar screen is available in widths of 2 to 4 feet and is inclined at 30 degrees.

FRONT CLEANED, REAR RETURN BAR SCREENS

The front cleaned, rear return Bar Screen is considered a "heavy duty" screen that can be used in channels over 70 feet deep and 12 feet wide. It consists of multiple rakes mounted on two strands of chain that ascend to clean the front side of the bar rack, and return on the rear, downstream side of the bar rack.

A pivoting deflector plate is required at the bottom of the screen to prevent debris from jamming in the boot area as the rakes revolve around the footsprockets. The deflector plate is pivoted out of the way by an ascending rake, and closes by gravity as the rake passes.

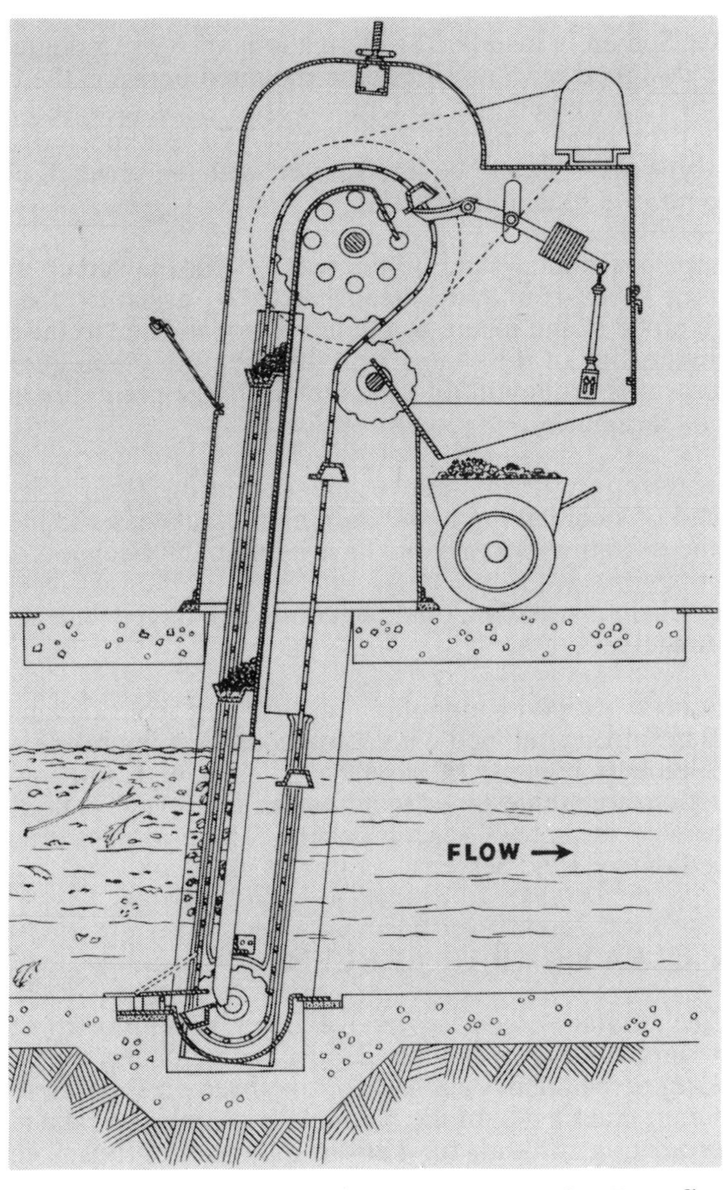

Front Cleaned-Rear Return Multiple Rake Bar Screen
Courtesy of Envirex, Inc.

This Bar Screen is installed at inclinations of 6 to 15 degrees, and may be designed with its sideframe mounted between the channel walls, or recessed in the concrete walls. If a recessed boot is provided, or large amounts of sand or grit is expected, special bucket-type rakes should be interspersed with the standard cleaning rakes to remove settleable material.

Discharge is accomplished on the rear side of the screen after the ascending rakes round the headshaft. To assist in the debris removal process, and insure that debris is not allowed to fall into the downstream side of the channel, an idler shaft or chain guide track arrangement is provided to locate the discharge point directly over the debris receptacle.

The headsection frame above the operating floor should be fabricated of steel having a 1/4" minimum thickness and the frame below the operating floor should have a minimum thickness of 3/8".

The rake chain is usually a pintle or combination type with a pitch of approximately 6".

Fine Screen Several manufacturers offer a front cleaned, rear return Bar Screen equipped with continuous cleaning rake elements. These elements may be of stainless steel or high impact plastic construction, are are designed to automatically "eject" debris as they rotate around the screen headsprockets. These screens are further discussed in the "Fine Screens" section of this book.

BACK CLEANED BAR SCREENS

Back Cleaned Bar Screens are available in rake widths from 2 to 10 feet, for applications with water depths generally less than 12 feet. The raking mechanism of a back cleaned screen is located on the downstream (back) side of the bar rack, with only the rake teeth, or tines, projecting through the bars in the cleaning position. This arrangement has the advantage of minimizing the possibility of debris jamming the rake mechanism, or wedging between the bar rack.

The rakes are chain mounted, and revolve over head and

footsprockets. As the descending chains revolve around the footsprockets, the rake teeth penetrate the bar rack from the downstream side. The teeth project through the upstream side of the rack by 4 to 7 inches, collecting debris as the ascend.

The bars are anchored at the channel bottom only, and are supported and spaced by the ascending rakes. The rakes "combing" action through the bar rack prevents the use of fixed, intermediate horizontal bar supports, and limits the effective depth of the screen.

If debris is not effectively removed by the wiper, or other cleaning mechanism, it can be carried over to the downstream side of the screen. Heavy accumulations of grit at the base of the bar rack can cause the screen to jam.

Vertical Back Cleaned Screens Vertical Back Cleaned Bar Screens require very little floor space and channel area, and are frequently used in high-lift pump stations and deep sewers where there are considerable differences between the operating floor and channel invert elevations. This screen has a maximum effective bar rack depth of approximately 12 feet, but may provide a vertical lift of up to 60 feet.

The rakes are spaced at 6 to 8 foot intervals and also serve as horizontal bar spacers. Most screens have a pivoting spacer mechanism located near the top of the bar rack that helps maintain proper bar spacing. The bar spacer is pivoted out of position each time an ascending rake passes. After removing debris from the bars, the rakes act as elevators to lift debris to the discharge point.

Although chain tracks are provided for the ascending and descending chain runs, this screen requires a minimum frame support structure. The cost for adding an additional foot to the depth of the screen is usually limited to the cost of 4 addtional feet of chain.

Debris is removed on the front, upstream side of the screen by an automatic rake wiper. A pivoting deadplate and debris chute are require to assure that debris does not fall back into the channel.

Cast polyurethane rake assemblies were initially designed for vertical Back Cleaned Screens, and the Vertical Back Cleaned

Screen may benefit most from their use. The reduced coefficient of friction of the polyurethane formulation used, its reduced weight, and the fact that it is slightly pliable are all seen as advantages in this application.

Inclined Back Cleaned Screens Inclined Back Cleaned Screens are furnished with the bar rack positioned at an angle of 9 to 30 degrees from vertical. Unlike vertical Back Cleaned Screens, these units discharge on the downstream side of the screen.

Front View of Back Cleaned Bar Screen
Courtesy of Hubert Water Systems BV

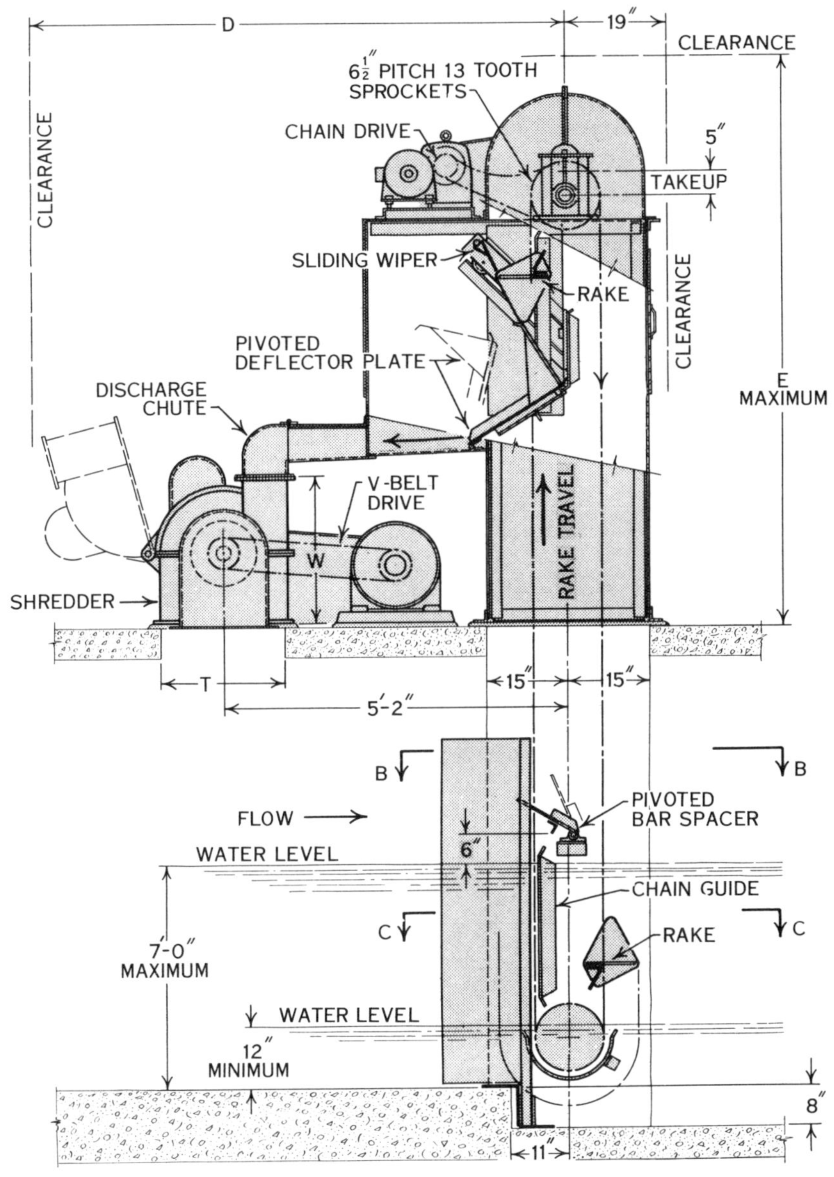

Back Cleaned Multiple Rake Bar Screen
Courtesy of FMC Corporation

Back Cleaned Multiple Rake Bar Screen
Courtesy of Electromecanique Verbandt Andre

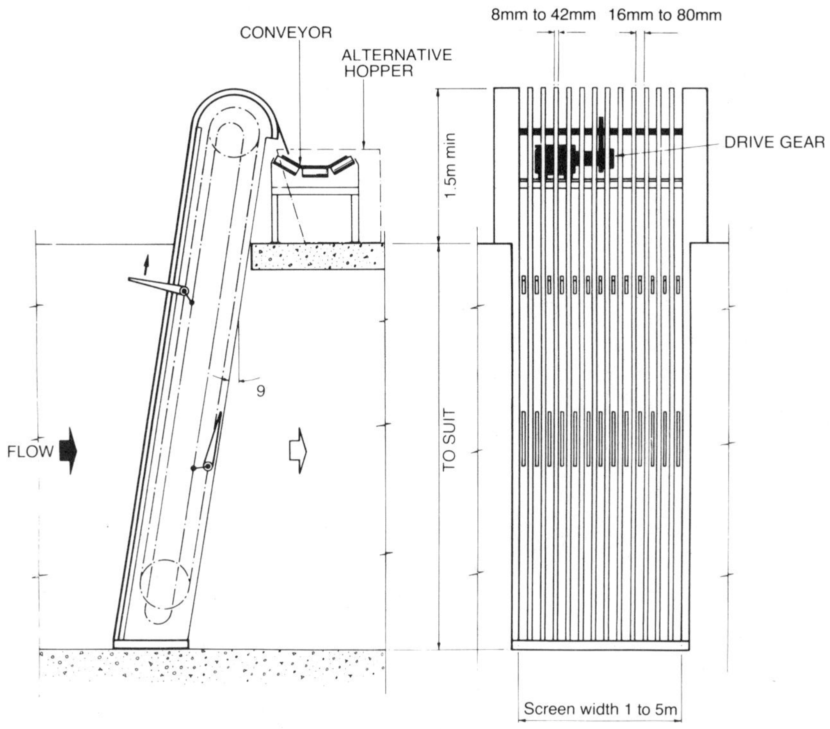

Back Cleaned Multiple Rake Bar Screen
Courtesy of Electromecanique Verbandt Andre

This screen is generally considered a "heavier duty" design than a vertical screen and its inclination allows it to handle more debris.

Some designs are equipped with bars that extend from the channel bottom, up the front of the screen, and over the headshaft to the discharge position. This design helps facilitate debris removal and reduce carryover by stripping debris from the rakes.

These screens are furnished with a self-supporting frame, and include guide tracks for the ascending and descending chains. Bar racks may be fabricated of rectangular or round bars.

Inclined Back Cleaned Bar Screen
Courtesy of Jeffrey Division, Dresser Industries

CATENARY BAR SCREENS

Catenary Bar Screens were developed for unattended use at large storm water pumping and drainage projects where the debris loads may be heavy or varied, and screen use may be infrequent. Catenary Bar Screens are now being applied on an increasingly frequent basis at municipal and industrial sewage treatment plants.

Like other multi-rake screens the Catenary Bar Screen consists of chain mounted toothed rakes. The most significant difference is that

Single Shaft Catenary Bar Screen
Courtesy of Jeffrey Division, Dresser Industries

the Catenary Screen has no footshaft or lower chain guide assembly. The screen chain descends into the channel to form a "catenary" curve near the channel bottom.

The underwater path of the chain and rakes is unguided, allowing them to ride over a surge accumulation of debris without jamming or overloading. The rakes are weighted to bear down against the face of the bar rack and collect and retain debris as they ascend to a discharge point above the operating floor. Rake cleaning is accomplished by a pivoting rake wiper mechanism which wipes debris from the rake into a suitable receptacle.

Combination type chain having 6.05" pitch or pintle chain with a 6" pitch are most frequently used for these screens.

The ends of each rake may be equipped with a replaceable, hardened steel or UHMW polyethylene wear pad. This pad engages the bars and/or deadplate on the rakes ascent and minimizes contact and subsequent wear of the rake teeth and screen bars.

Double Shaft Catenary Bar Screen
Courtesy of Jeffrey Division, Dresser Industries

Catenary Bar Screens are available in widths ranging from 2 to 30 feet, and depths to 50 feet or more. Bar rack inclinations of 15 to 45 degrees are recommended. The catenary loop of the chain may prevent the bottom 4 to 10 inches of the bar rack from being effectively cleaned. It is recommended that the lowest portion of the bar rack be curved into the flow to prevent debris accumulations at the channel bottom.

Units are available with single or double shaft designs. Double shaft screens are recommended for heavier duty applications, and those installations with bar rack inclinations of 30 degrees or more.

The structural support for a Catenary Screen drive mechanism consists of a relatively simple headframe assembly that is anchored to the operating deck. The descending, return run of the chain is guided by runway angles bolted to the channel walls that may extend to within 3 feet from the channel bottom.

Counterweighted Rakes on a Catenary Bar Screen
Courtesy of Fairfield Service Company

RECIPROCATING RAKE BAR SCREENS

Reciprocating Rake Bar Screens are automatic bar screening devices that use a single, reciprocating rake to clean the stationary bar rack. The reciprocating action of the rake simulates manual raking, and reduces the possibility of jamming the rake mechanism. Several versions of these screens have been widely used in Europe for over twenty years, and they are being specified on an increasingly frequent basis in the United States.

The increasing popularity of these screens is due to the fact that they have relatively few mechanical parts, no moving parts permanently located below the designated maximum water level, and most inspection and maintenance can be done from the operating floor level. Because so few parts are exposed to the water, it may be economically feasible to require that all "wetted" parts be constructed of stainless steel or other suitable corrosion resistant material.

The operating cycle of a reciprocating rake screen begins with the rake in the standby, or parked position at the discharge elevation. The cleaning rake is lowered in the channel with the rake teeth in an extended, disengaged position. The cleaning rake is fixed to the end of a long arm that projects down into the channel, allowing the operating mechanism to remain above the waters surface. As the rake reaches the channel bottom, the teeth are rotated, or pivoted into an engaged position with the bar rack, and the rake is lifted up the face of the screen. The screenings are removed from the rake into a suitable trash receptacle by an pivoting wiper mechanism.

The typical rake travel speed for Reciprocating Rake Bar Screens is 20 feet per minute. Some designs are able to use a two speed drive motor that can provide a higher speed (i.e. 40 fpm) for the descending stroke of the rake. This arrangement will produce increased debris handling capability by adding a significant number of rake cleaning cycles per hour. The drive motor should be equipped with an integral motor brake to allow the rake carriage to be stopped at any operating level.

Although Reciprocating Rake Bar Screens have been manufactured with effective rake widths of 22 feet, they are generally considered for applications 2 to 12 feet wide, and depths up to 25 feet (It should be noted that actual depth restrictions do vary among manufacturers). The 25 foot maximum depth is recommended for most applications because the length of time required to complete one cleaning cycle may be prohibitive on deeper installations.

Like the drive components, most of the screen framework is located above the water level, and the frame is usually anchored to the operating deck. The bar racks for Reciprocating Rake Bar Screens may be mounted at a 0 to 25 degree inclination. Most reciprocating rake screens are of the front cleaned design, eliminating the problem of debris carryover although several manufacturers also offer back cleaned versions of this screen.

Disadvantages of the reciprocating rake design include the debris handling limitations caused by having only one cleaning rake, and the high headroom requirements of some screen designs.

The major difference among the Reciprocating Rake Bar Screen designs that are currently available is the drive mechanism that is used to accomplish the raising and lowering of the cleaning rake. Each manufacturers' design has unique features that, if specified, can exclude competitive designs.

Cable Operated Reciprocating Rake The Cable Operated Reciprocating Rake Bar Screen is one of the oldest types of mechanically cleaned screens. These screens utilize a carriage-mounted rake assembly that is raised and lowered by one, two, or four stainless steel cables, or "ropes", to clean the stationary bar rack. The cables operate from grooved drums mounted on a common headshaft. Rake travel drums are keyed to the shaft to control the up and down movement of the rake, and rake tooth positioning drums are bronze bushed to operate freely. Controlled slippage between the drums results in the pivoting of the rake teeth.

The rake carriage is equipped with guide blocks that travel vertically within channel guides embedded in the the walls of the concrete channel to assure positive rake engagement.

A pivoting rake wiper may be mounted on the screen headframe, or furnished as an integral part of the rake carriage mechanism.

Cable operated screens are available in widths of up to 20 feet, and depths of up to 350 feet.

Cable Operated Reciprocating Rake Bar Screen
Courtesy of FMC Corporation

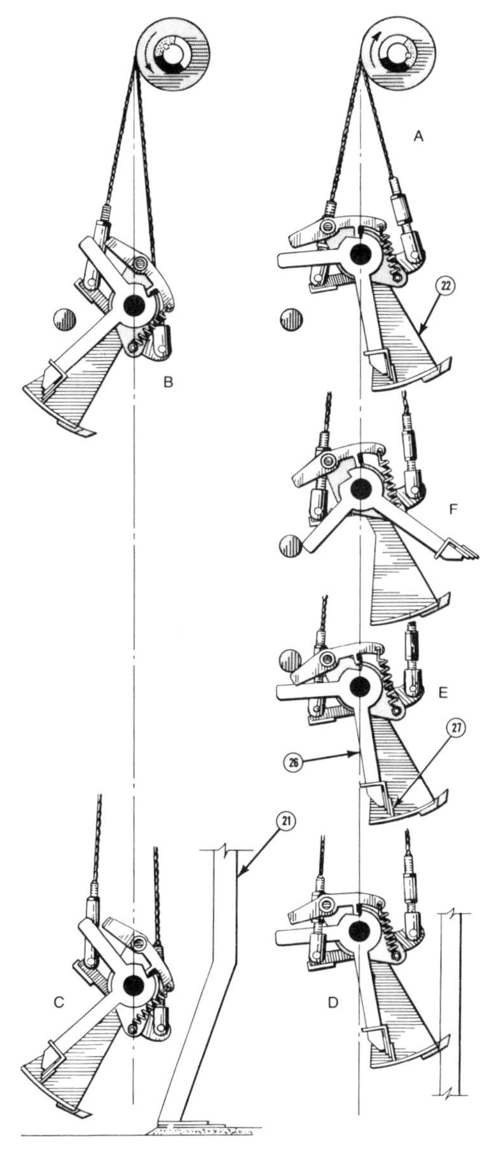

Operational Diagram of Cable Operated Bar Screen
Courtesy of FMC Corporation

Inclined Cable Operated Bar Screen
Courtesy of FMC Corporation

Cogwheel Driven Reciprocating Rake Some of the most successful and versatile reciprocating rake designs utilize a cogwheel-type drive arrangement to raise and lower the cleaning rake. On these designs, the entire cleaning rake assembly, including the gearmotor, is carriage mounted on a pair of cogwheels that travels around, or up and down, a fixed rack. The design of the fixed rack on which the cogwheels travel varies among screen manufacturers. Pin rack, toothed track rail, and gear rack designs are available.

The engagement and disengagement of the rake teeth and screen bars is controlled by the position of the cogwheels on the fixed rack. As the cogwheel travels around the bottom of the rack, the cleaning rake is rotated into the engaged position for the ascending, cleaning stroke. As the cogwheel travels around the top of the fixed rack, the rake is disengaged for its descending, return stroke. The rake arm is equipped with guide, or follower rollers that travel within a guide track to fix the upper position of the rake arm relative to the screen bars.

If the rake encounters an obstruction during the cleaning stroke, the cogwheels stop turning and the upward travel of the rake carriage is halted. The motor continues to operate, resulting in a tilting of the drive unit as it rotates slightly in a counterclockwise direction on the shaft. The tilting action compresses a coil spring and the rotating movement causes the rake to disengage. The carriage can again begin its upward movement, riding over the obstruction. If the object still interferes with travel and the teeth are in the disengaged position, the coil spring is further compressed, activating a limit switch, that stops the drive motor.

Headroom requirements for this screen design should be carefully considered. An estimate of the headroom that will be required can be made by adding the vertical discharge height above the operating floor to the vertical depth of the bar rack. This sum will approximately equal the minimum headroom that should be provided above the discharge elevation.

Multiple Reciprocating Rake Bar Screen Installation
Courtesy of Infilco Degremont, Inc.

Pin Rack Type Reciprocating Rake Bar Screen
Courtesy of Infilco Degremont, Inc.

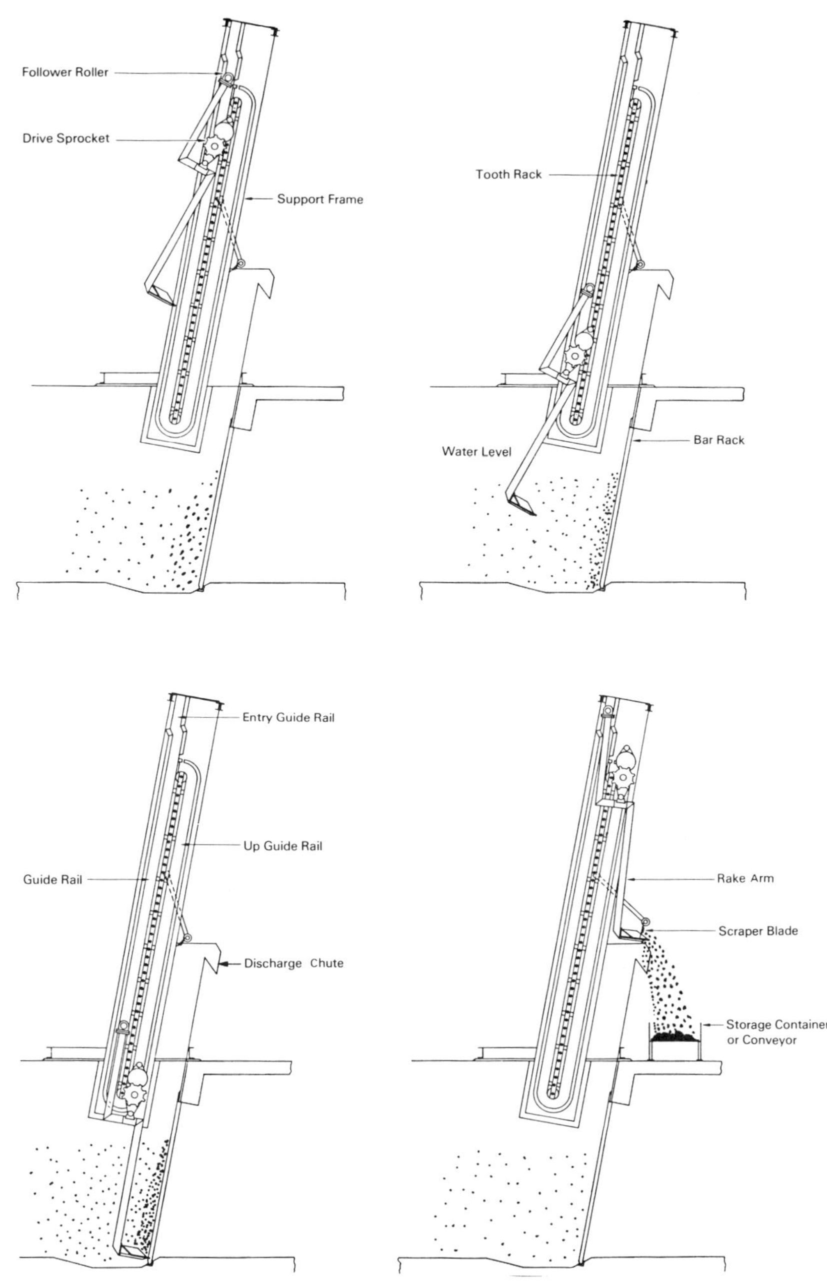

Operational Diagram of Reciprocating Rake Bar Screen
Courtesy of Infilco Degremont, Inc.

Cog-Rail Type Reciprocating Rake Bar Screen
Courtesy of Hycor Corporation

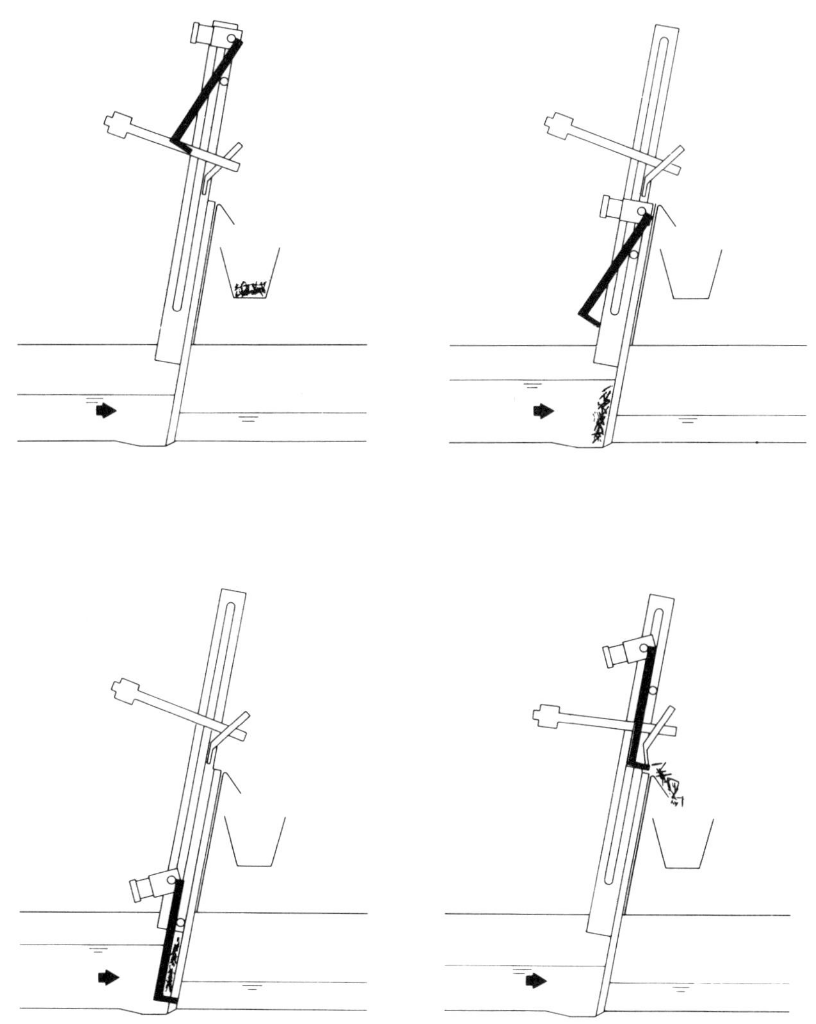

Operational Diagram of Reciprocating Rake Bar Screen
Courtesy of Hycor Corporation

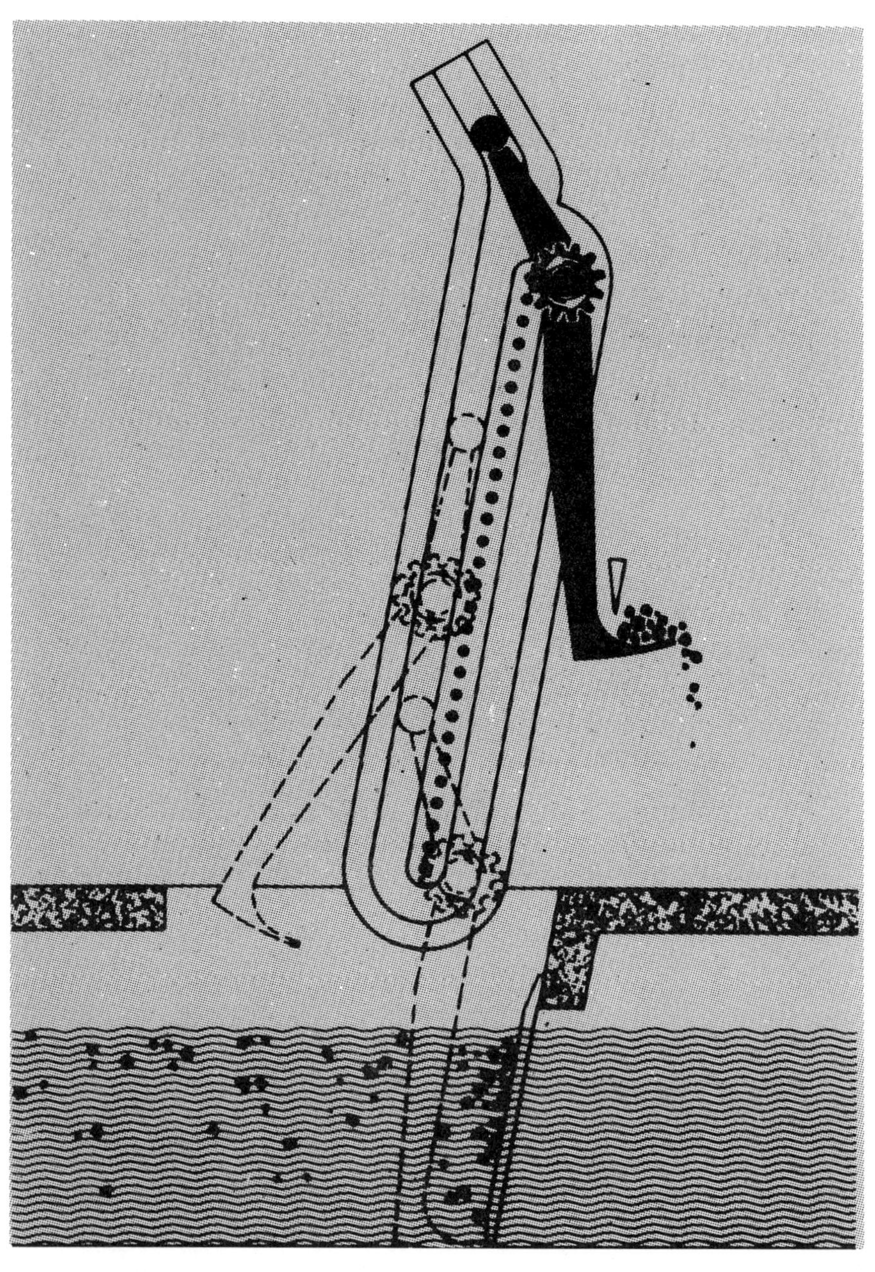

Operational Diagram of Reciprocating Rake Bar Screen
Courtesy of Hawker Siddeley Brackett

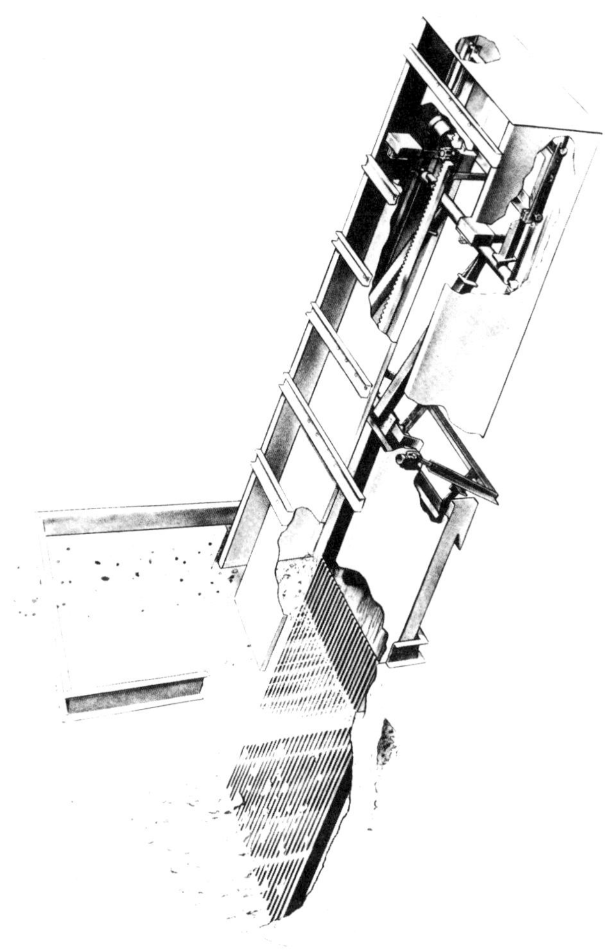

Reciprocating Rake Bar Screen
Courtesy of Jeffrey Division, Dresser Industries

Chain Lift Reciprocating Rake Operation of the cleaning rake may be accomplished through the use of two matched strands of 4" pitch, non-lubricated roller chain operating in guide tracks. The lower end of each strand of chain is attached directly to the rake

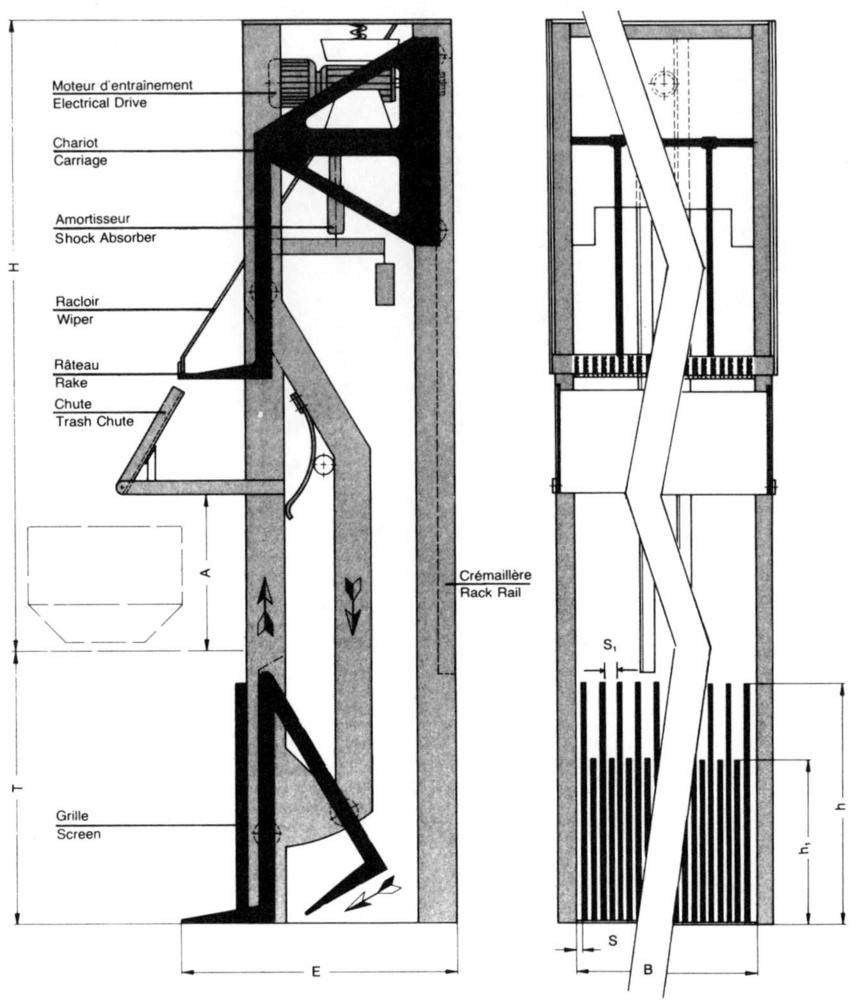

Back Cleaned Reciprocating Rake Bar Screen
Courtesy of John Meunier, Inc.

arms, and the upper end is fixed to screen headframe. The raising and lowering of the rake carriage is controlled by the operation of the roller chain over headsprockets mounted on a drive shaft in the screen headframe.

The close fitting chain guide tracks prevent the chain from buckling during their downward stroke and permit considerable power to be transmitted through the chain, driving the rake through most debris accumulations.

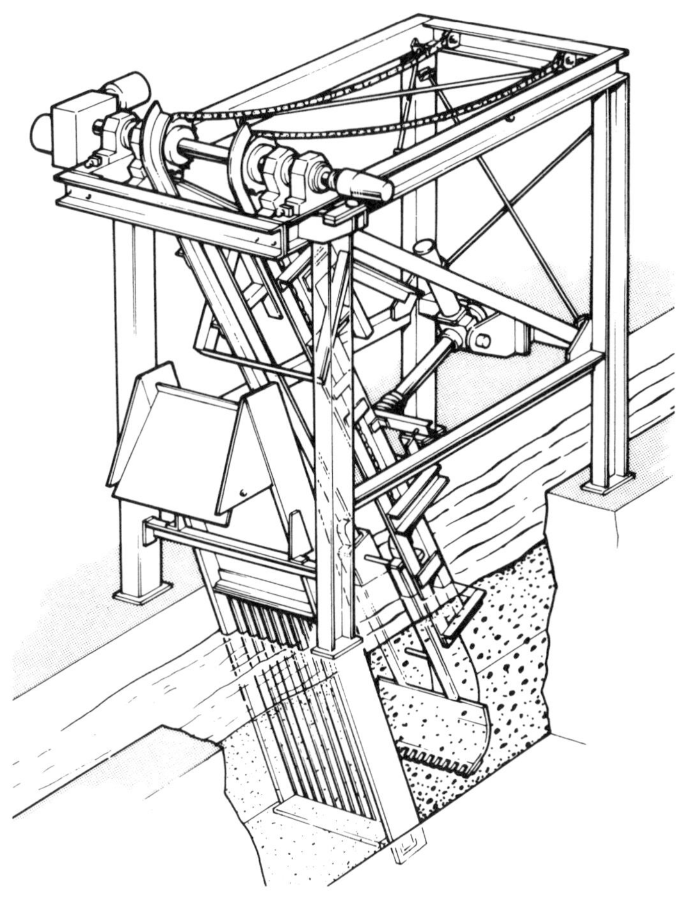

Chain Lift Reciprocating Rake Bar Screen
Courtesy of William Green, Ltd.

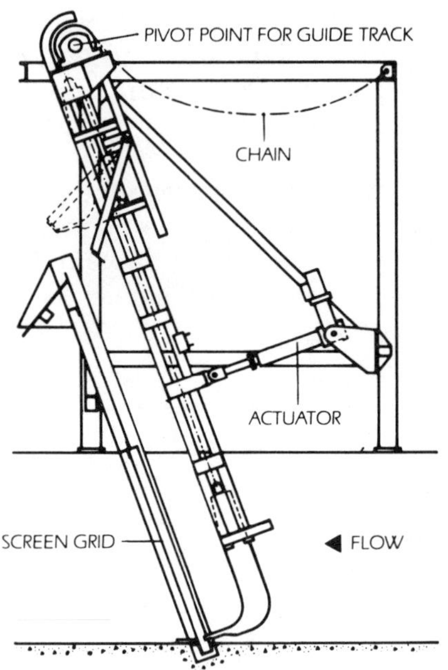

Chain Lift Reciprocating Rake Bar Screen
Courtesy of William Green, Ltd.

The chain/rake guide tracks are constructed in a rigid frame assembly, supported by the screen drive shaft, that allows the entire guide frame to be pivoted to engage or disengage the rake teeth with the bar rack. This pivoting movement is controlled by an electric actuator with a fractional horsepower drive motor that "powers" the rake into position.

A rotary limit switch on the screen drive shaft limits the up and down movement of the rake, and a cam type limit switch on the actuator drive controls the rake engagement and disengagement.

The screen is protected against rake overloads through the use of current sensing devices in the circuits of the screen drive and actuator motors. If the rake encounters an obstruction in any mode of operation (e.g. rake engage/disengage or rake ascent/descent), it is immediately recognized by the sensors as an increase in amperage, and translated into a retraction movement of the rake. As the rake clears the obstruction, the subsequent lowering of current initiates another retraction movement, and the rake continues operation at the point that it was interrupted. The rakes ability to "sense" an obstruction and continue to operate around it is termed "profiling".

Reciprocating Rake Bar Screens using a fixed gear rack may require substantial headroom to allow the rack to project above the screen discharge point. The chain lift design has the advantage of requiring less headroom than fixed rack designs. Excess chain that accumulates during the raising of the rake is stored on chain collecting bars located within the screen headframe. The maximum headroom required is approximately equal to the rake discharge height plus 6'-6".

This drive arrangement also allows the screens drive unit to remain in a fixed position on the headframe as the rake is raised and lowered.

Chain Driven Reciprocating Rake The cleaning rake of a reciprocating rake screen may be driven by two endless strands of roller chain operating over head and footsprockets. The rake is attached to the screen chain and, as the chain revolves around the sprockets, the rake moves in and out of the channel to clean the bar rack.

Each end of the rake is equipped with guide rollers that travel in guide channels to maintain the pivot point of the rake arm relative to bar rack. The rake/chain connection is spring loaded to allow the rake to pass over any lodged objects or obstructions that it encounters.

Screen chain may be furnished with a chain pitch of 1-1/4" to 6".

Chain driven rake screens have drive units located in a fixed position on the headframe.

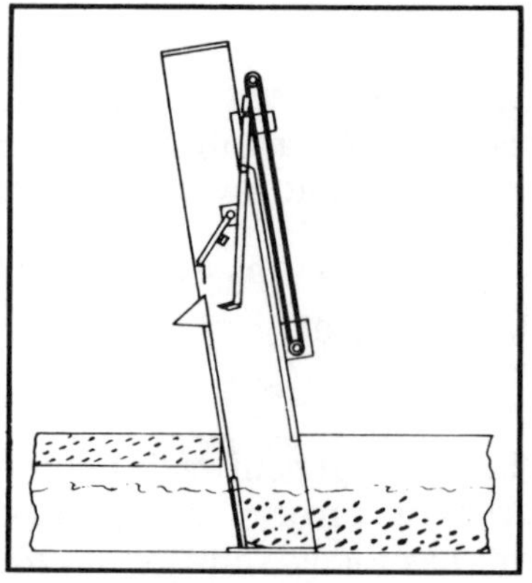

Normal rest position
The screening cycle begins when the rake mechanism is activated. The rake moves down the guides in the "pre-cleaning" position.

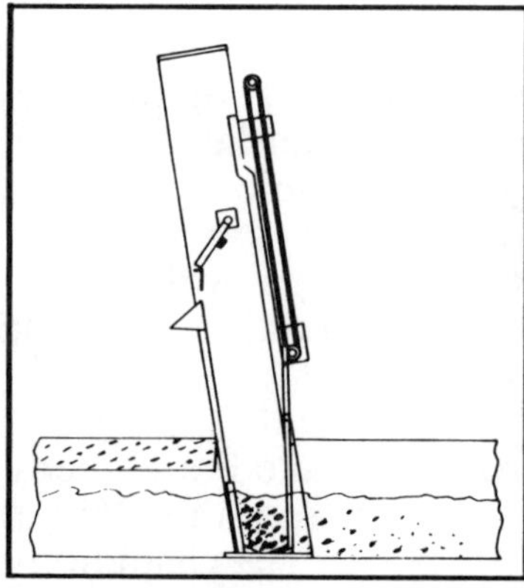

Downward cycle
The rake enters the water on the upstream side of the screen prior to cleaning the bars.

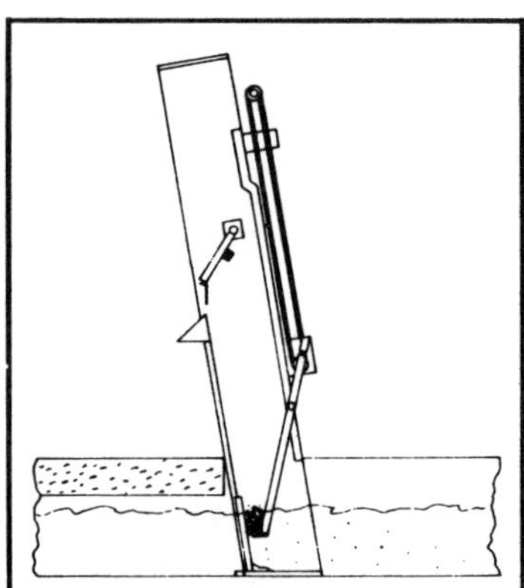

Cleaning of bars
As the mechanism rotates around the lower sprocket, the rake arm dynamically engages the rake head into the bar rack.

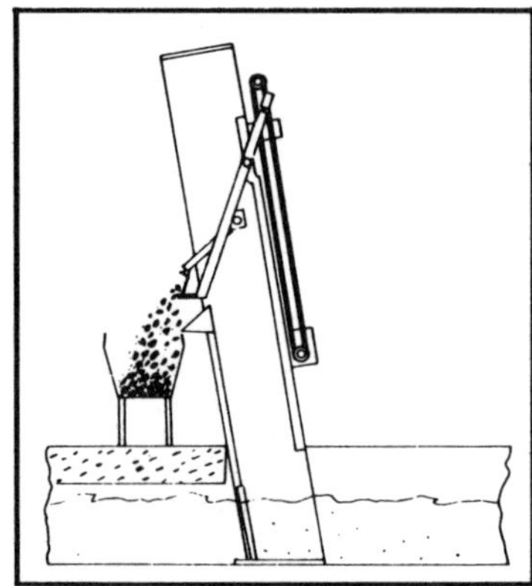

Upward cycle
The rake is lifted back up the guides delivering the screenings to the discharge chute where the rake head is positively cleaned by the wiper.

Operational Diagram of Reciprocating Rake Bar Screen
Courtesy of Vulcan Industries

Installation of Reciprocating Rake Bar Screen
Courtesy of Vulcan Industries

Hydraulically Driven Reciprocating Rake The raising/lowering and engage/disengage motions of a reciprocating rake screen can be provided through the use of hydraulic cylinders. This screen design is generally considered for light duty applications with channels less than six feet deep and four feet wide. The use of a cable-type lift cylinder or an auxiliary debris conveyors may allow use on deeper installations.

The cleaning stroke begins as a hydraulic cylinder mounted horizontally between the headframe and rake arm is retracted, pivoting the cleaning rake into the bar rack. A second hydraulic cylinder mounted to the top of the screen headframe is used to raise

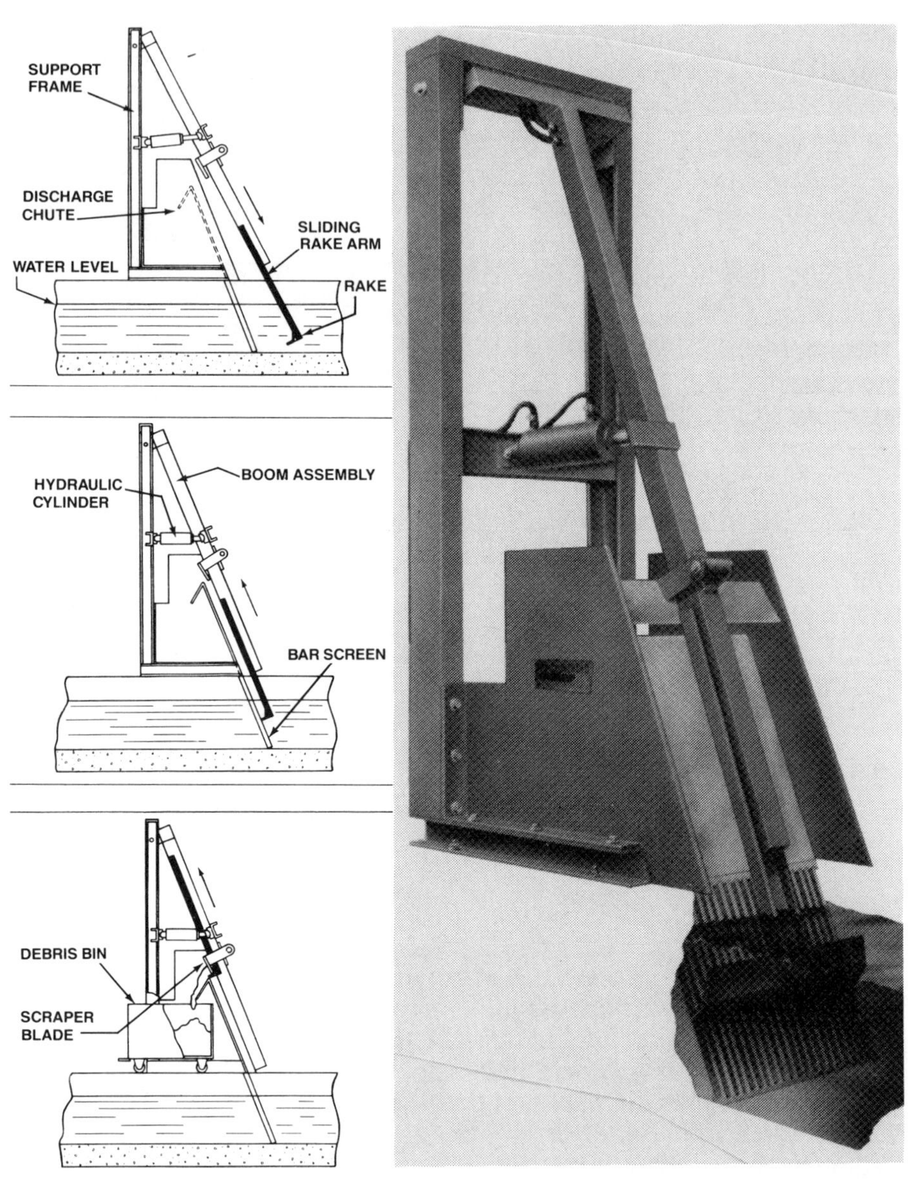

Hydraulically Operated Reciprocating Rake Bar Screen
Courtesy of Franklin Miller

and lower the cleaning rake. This cylinder may be a rod or cable type.

The synchronized motion of the cylinders is powered by a small hydraulic drive unit. Pressure relief valves provide overload protection to prevent equipment damage.

Hydraulically Cleaned Reciprocating Rake Bar Screen
Unknown Manufacturer

Screw Operated Reciprocating Rake This screw operated bar screen is a relatively light duty screen designed for use in small treatment plants.

The up and down rake movement is provided by a ball bearing nut that travels the length of a rotating screw. The vertically positioned screw is directly coupled to a reversing drive motor and operates in stationary upper and lower bearings mounted to the screen headframe. The rake carriage is fitted with a ball bearing drive nut that traverses the length of the screw as it is rotated.

Bar Screen with Angled Bar Rack
Courtesy of Enviro-Care, Inc.

Screw Operated Reciprocating Rake Bar Screen
Courtesy of Enviro-Care, Inc.

Rake arms project down from the rake carriage and are equipped with upper and lower guide rollers. The rollers travel in guide channels to provide tooth engagement on the rakes ascending stroke, and disengagement on the descending stroke.

Upper and lower proximity switches limit the travel of the rake carriage.

ARC SCREENS

An Arc Screen is a self-cleaning bar screen consisting of a curved bar rack that is cleaned by one or more revolving rake assemblies. These screens are designed for applications in small plants with channel depths usually less than 7'-0", and they are available with a variety of rake configurations.

The raking mechanism is mounted on a drive shaft that operates in pillow block bearings and slowly rotates about a horizontal axis. The rotation of the cleaning rake cleans the radius of the stationary bar rack and elevates debris to a discharge point at the top of the rack, where it is cleaned by a wiper mechanism. Some units are provided with a pivoting debris plate to prevent debris from falling back into the channel.

The rake teeth fully engage the bar rack and may be furnished with an optional spring loaded mechanism to override minor obstacles. The bar rack is constructed of steel bars formed in a radius that is approximately equal to the sum of the channel depth and the rake discharge height.

Full Rotary Arc Screen
Courtesy of Biwater, Ltd.

Arc Screens may utilize direct coupled, chain and sprocket, v-belt, or shaft mounted drive units. Overload protection can be provided by a torque limiting coupling, shear pin device, current sensor, or torque overload switch.

Full Rotary Arc Screens The cleaning stroke of the Full Rotary Arc Screen is a complete 360 degree circle and its cleaning mechanism is very simple and reliable. Screens may be fitted with one, two, or four cleaning rakes.

Screens may be installed in concrete channels for through-flow operation, or above floor openings and drains for down-flow applications.

Screens up to 4 feet wide may be installed in concrete channels, and widths up to 12 feet are available for down-flow screens having 3 to 6 foot rake diameters.

Full Rotary Arc Screen
Courtesy of Infilco Degremont, Inc.

Semi-Rotary Arc Screens The rotary motion of the screen drive unit can be converted into a semi-rotary action by means of a simple pivoting linkage. As the rake completes its upward cleaning stroke, it disengages and returns to the bottom of the curved bar rack, re-engages, and repeats the cycle.

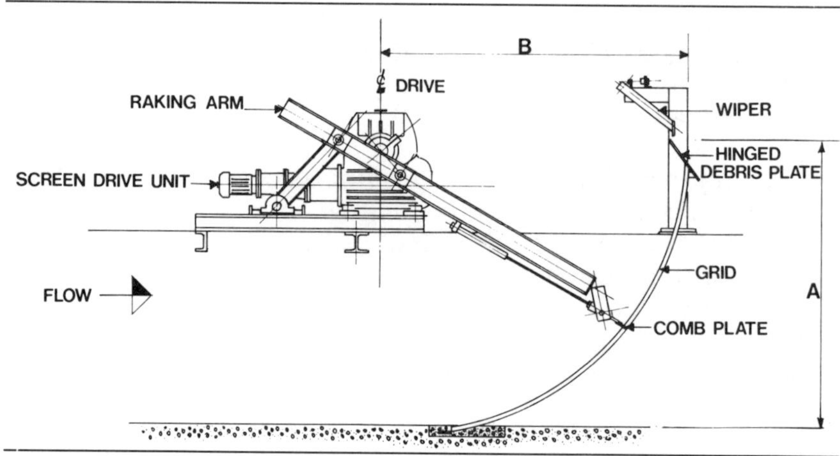

'A' Channel Invert to discharge point (M)	1000	1250	1500	1750	2000	2250	2500
'B' Minimum length of channel (M)	2000	2300	2750	3100	3500	3800	4100

Semi-Rotary Arc Screen
Courtesy of William Green, Ltd.

The Semi-Rotary Arc Screen is furnished with a single cleaning rake. Its primary advantage is that it requires approximately one half of the channel length and head room required by a full rotary screen.

OTHER BAR SCREENING EQUIPMENT

MANUALLY CLEANED BAR RACKS

Manually cleaned bar racks are frequently the only type of screening employed in small treatment plants. Larger plants that utilize mechanically cleaned Bar Screens for primary screening purposes may also utilize manually cleaned bar racks in overflow bypass channels.

Manually cleaned bar racks should be limited in depth to that which can be conveniently raked by hand. The racks are normally placed at an inclination of 30 to 45 degrees from horizontal to increase available screen area by 40 to 100 percent. This inclination will also assist the operator in his ability to drag up large debris with the manual rake.

Bar spacings for manual rakes are typically 1 to 2 inches, and the size of a manually cleaned rack usually measures approximately 20 square feet of screening area per MGD. A horizontal, perforated drainage plate may be furnished at the top of the rack to temporarily store debris for drainage.

Manually cleaning may be required two to five times per day. Debris buildup on the racks must be carefully monitored to insure reasonably free flow into the treatment plant.

COMMINUTING MACHINES

The Comminutor is an in-channel screening device that is used to continuously screen large solids and comminute, or cut, them into smaller, settleable particles. Reduced comminuted particles are more easily transferred without clogging pumping equipment, and are more suitable for the digestion process.

A Comminutor consists of a slotted drum that slowly rotates on a vertical axis. The drum is a combination screen and support casting for the peripheral cutting teeth and shear bars. As the drum rotates, solids too large to pass through the screen slots are cut by the cutting teeth until they are small enough to pass through the slot openings.

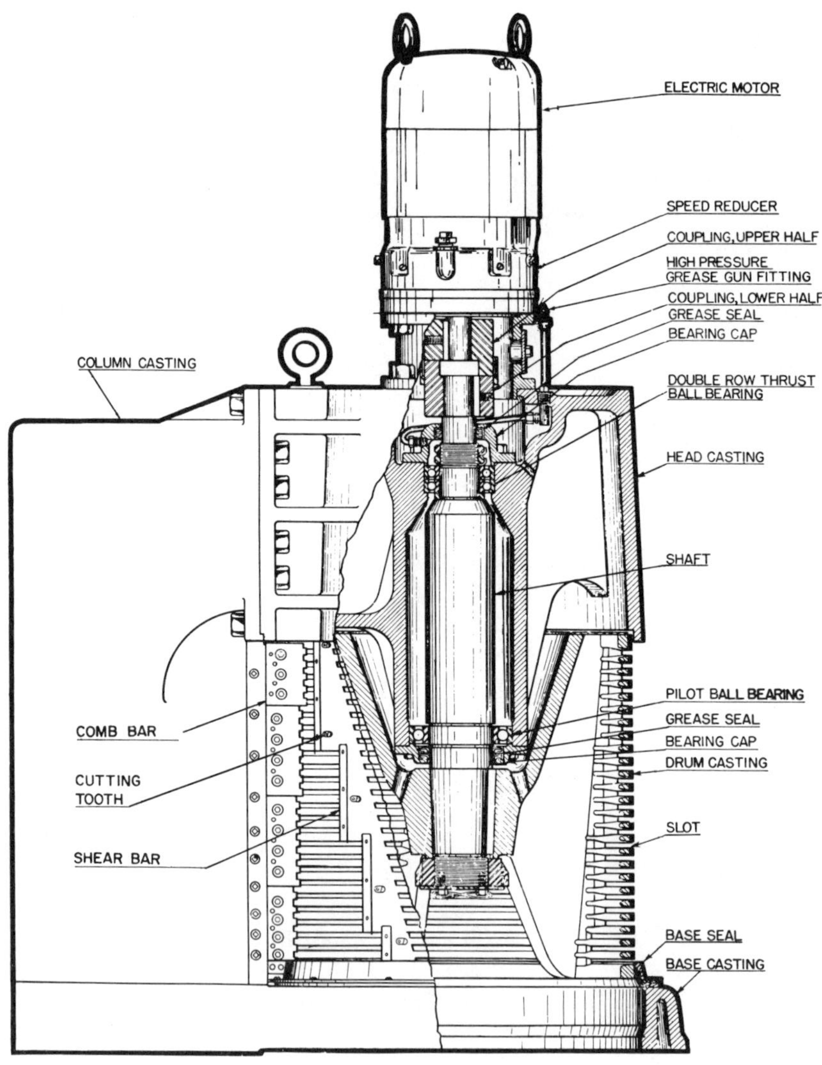

Vertical Shaft Comminutor Sectional Diagram
Courtesy of Chicago Pump Company

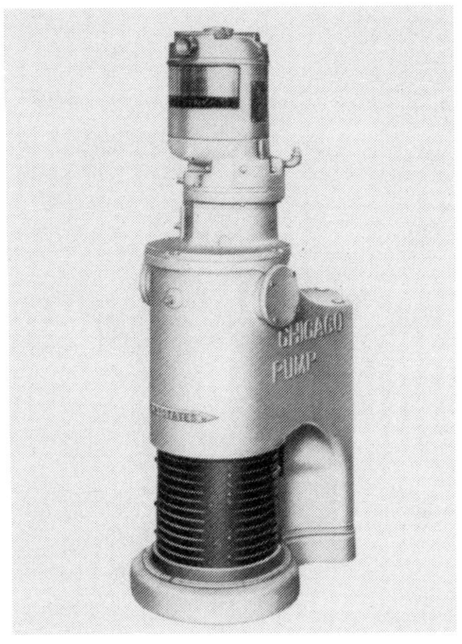

Vertical Shaft Comminutor
Courtesy of Chicago Pump Company

Another Comminutor utilizes a conical screen rotor carried by a central shaft and rotating about its horizontal axis. Hardened steel cutters are mounted around the periphery of the rotating screen to provide a cutting action against stationary knives fixed to the machine frame. This machine can be hydraulically or electrically driven.

Comminutors may be installed in lift stations, sewage treatment plants, and industrial applications. Because their cutting teeth are not able to cut hard objects including stone and metal particles, Comminutors are usually not recommended for combined sewer flows.

Comminutors are available with slot widths of 1/4" to 3/8", and drum diameters of 7" to 54". Gearmotor horsepowers typically range from 1/4 hp on small units to 5 hp on larger units.

Horizontal Shaft Comminutor
Courtesy of Enviro-Care, Inc.

At least one manufacturer offers a combination bar screen and comminuting machine with a rotating cutter that travels up and down the face of a stationary bar rack. The cutter rides freely on guide rails and can swing away from the screen if non-comminutable solids are encountered. The cutter is raised and lowered using a wire rope and drum, or telescoping screw arrangement. Rotation of the cutting unit is alternated to prevent debris from accumulating. This unit is available with 3/8" to 3/4" bar spacings for channels 1 to 8 feet wide. On 1 to 3 foot wide machines a single, 2 hp to 3 hp motor drives the rotating cutter and the telescoping screw arrangement. On units 4 to 8 feet wide, in addition to the 5 hp to 10 hp comminuting motor, a 1 hp hoist motor is provided to raise and lower the cutter unit.

Combination Bar Screen and Comminutor
Courtesy of Chicago Pump Company

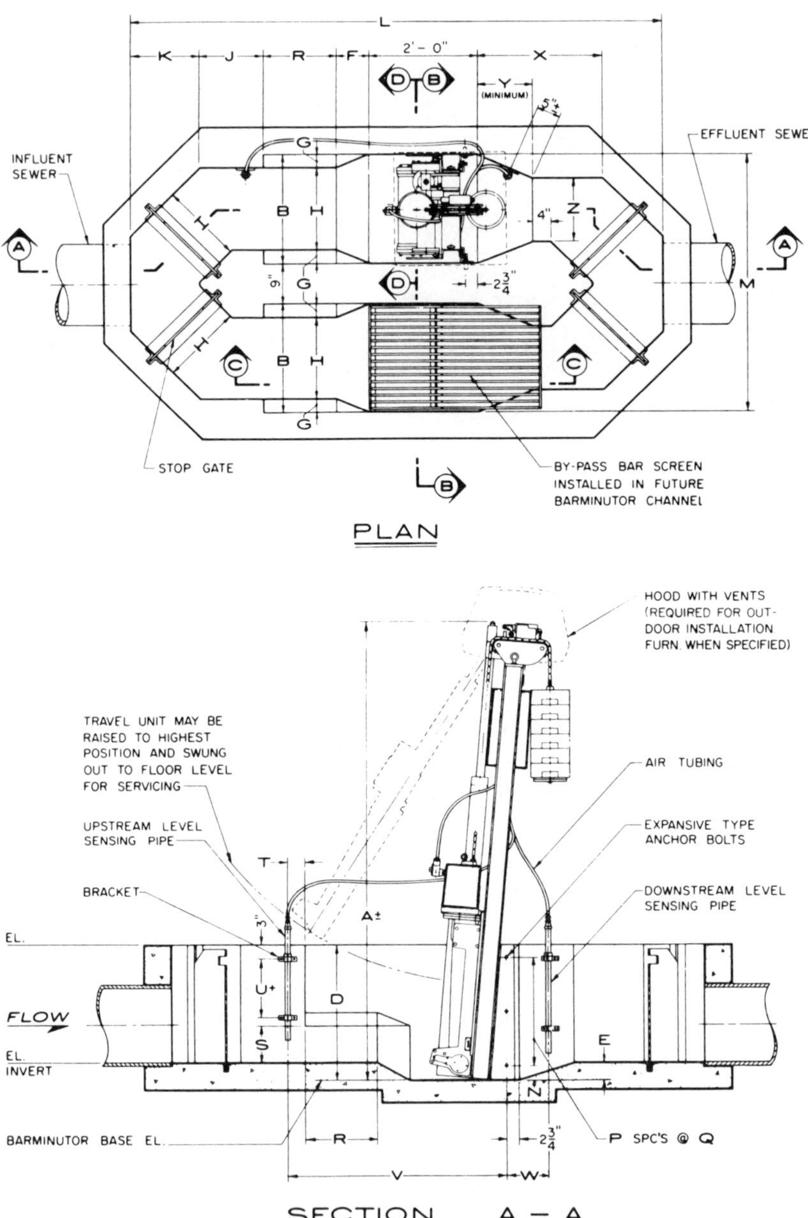

Suggested Layout of Bar Screen/Comminutor

Courtesy of Chicago Pump Company

BAR SCREEN COMPONENTS

BAR RACK

The common component of all Bar Screens, whether mechanically or manually cleaned, is the bar rack. The bar rack, sometimes called the bar grid, is a stationary grate constructed of equally spaced vertical bars. Uniform bar spacing is be maintained through the use of horizontal support members with milled slots or welded spacers to which the vertical bars are attached.

Bar racks typically extend from the bottom of the channel to a point at least 9" above the maximum water level where they are attached to a steel deadplate that extends to the debris discharge point. Many screens are designed to have their framework recessed in the concrete walls of the channel so that the entire width of the channel is available as effective screening area.

The bar rack should be structurally sound and capable of withstanding the loads imposed by a minimum headloss of two feet. Although many sizes and shapes of bars may be utilized, it is normally recommended that the section modulus be equal, or greater than that of a rectangular bar measuring 3/8" thick by 2" wide. The width, depth, bar spacing, and headloss requirements will determine the actual size requirements.

Debris often becomes wedged between rectangular bars, just beyond the reach of the cleaning rake tooth. Most manufacturers are now offering racks constructed of bars having a trapezoidal cross section to help avoid this problem. Trapezoidal bars are oriented on the rack using the base of the trapezoid as the upstream screening surface, with the bar thickness decreasing progressively in the direction of the flow. This prevents debris frome becoming wedged between the bars; the debris either remains matted on the face of the rack, or flows completely through it.

When specifying trapezoidal bars, it should be considered that some bar sizes may be non-standard and may not be stocked by all steel supply houses. Most manufacturers standardize and stock one or two bar sizes that they have purchased in large mill-run quantities.

If a particular cross section is specified, it may provide a significant price advantage to the manufacturer that stocks the specified size. If trapezoidal bars are preferred, it is recommended that a specification reference an approximate bar size and/or a minimum acceptable section modulus.

At least one manufacturer offers a bar rack constructed of rectangular bars positioned at an angle to the flow. This design may result in improved flow through the unit, and when used in conjunction with a saw-toothed rake design, provides full engagement of the cleaning rakes with the bars.

When possible, the bar rack should be constructed so that it can be readily removed, without disassembling the screen or removing it from the well. Some installations have been provided with a bar rack that can be automatically lifted from the channel during extreme flow or headloss conditions.

DEADPLATE

In most applications it is not necessary or economical to extend the bar rack to the discharge point. Most bar racks terminate at some point above the maximum water level and are attached to a deadplate that continues to the point of discharge.

Deadplates are normally constructed of 1/4" thick steel plate, and are fabricated with 6" high sideplates to prevent screenings from falling off the ends of the rakes. The dead plate should be constructed true and flat so that the clearance between the deadplate and rake teeth is approximately 3/16".

CLEANING RAKES

Mechanically Cleaned Bar Screens are equipped with cleaning rakes that have teeth cut to match the openings of the bar rack. The teeth on front cleaned designs should project at least 1/2" into the bar rack to effectively clean the front and sides of the stationary bars and provide a minimum lifting shelf of 5-1/2". The rakes on back cleaned designs should penetrate through the bar rack and provide a minimum lifting surface of 4" on the upstream side of the screen.

Rake teeth may be flame cut or milled out of 3/4" thick steel plate. For bar openings of less than one inch, it is recommended that segmented toothed rake sections be considered. These rake segments are 12" to 20" long, and are bolted directly to the rake body. They may be individually adjusted to minimize tooth/bar rack interference, and compensate for tolerances accumulated during fabrication of the bar rack. Cast iron or cast steel rake tooth segments are available from some manufacturers.

Several manufacturers offer rake designs that have individual toothed rake segments spring loaded to the rake body. This allows the rake to ride over minor obstacles without overloading the screen mechanism.

A manufacturer of elastomers has recently developed a cast urethane rake segment for use on both front and back cleaned Bar Screens. These toothed rake segments are bolted to a steel rake body and mounted on the screen chain resulting in an overall rake weight reduction of approximately 65% on some cleaning rake designs. The low coefficient of friction, and the abrasion resistance of the urethane material should provide increased performance.

Multi-Rake Designs Cleaning rakes are furnished in multiples of two and should be spaced at equal intervals ranging from 16" to 12'-0". The rakes are attached to the screen chain with stainless steel bolts at special chain attachment links.

Reciprocating Rake Designs The cleaning rakes on single, reciprocating rake screen designs should be able to collect and retain all the accumulated debris on the face of the bar rack in a single cleaning cycle. It is recommended that the cleaning rakes on these screen designs have minimum 8" wide lifting shelf, although some manufacturers offer rakes as wide as 14".

The screen mechanism and rakes should be designed to provide a minimum lifting capacity of 60 pounds per foot of rake width.

HOUSINGS AND SCREEN ENCLOSURES

Multi-Rake Screens With the exception of the Catenary Bar Screen, all multi-rake screens are furnished with a housing to fully or partially enclose the screen headsection. A housing is furnished as an option on Catenary Screens.

Standard housings are fabricated from #12 or #14 gauge steel, with stainless steel, aluminum, and fiberglass offered as options. Housings should provide easy access to the screen machinery for inspection and maintenance purposes. In addition to their aesthetic value, screen housings are important for safety and odor control purposes.

Reciprocating Rake Screens Most Reciprocating Rake Bar Screens are furnished with an expanded metal or wire mesh safety enclosure that surrounds the three open sides of the screen. This enclosure should extend to a height of 6'-6" above the operating floor. Complete enclosure of the screen above the operating floor is available as an option by most manufacturers.

Reciprocating Rake Screen with Guards
Courtesy of Passavant Corporation

MECHANICAL BAR SCREEN SIZING PROCEDURES

Proper selection and sizing of a Bar Screen will insure satisfactory mechanical performance and increase the efficiency of downstream equipment and processes in the treatment plant. The information typically used to determine the size of the required Bar Screen includes:

* *Maximum and average daily flow*
* *Minimum and maximum water levels*
* *Bar spacing*
* *Number of screens required*
* *Expected debris loading*
* *Description of upstream & downstream processes*
* *Method of debris disposal*

There are several rules-of-thumb that can be used for preliminary screen size selection: 20 square feet of screening area per mgd of sewage or, a clear area at least 50% greater than the cross section of the contributing sewer. However, the procedure that is recommended to be used to determine the size of a Bar Screen involves the determination of the size of the bar rack, plus suitable freeboard, to pass the required flow at a velocity that prevents sedimentation.

MODEL/TYPE OF SCREEN

Some types of Bar Screens are capable of handling a flow of over 150 million gallons per day per screen. It is therefore important that the application be carefully reviewed and all operating, installation and maintenance costs be considered. Channel width and depth, debris loading, climate, water chemistry, and lack of adequate maintenance may severely limit the choices of screen models available for a given application.

If an incorrect type of Bar Screen is selected or specified, or if the specifications are not strict enough, problems may result. Screens that are over and under-designed for an application both result in wasted money.

As with most types of mechanical equipment, maintenance requirements and reliability are important aspects of design. Some applications may require a screen to operate as infrequently as ten minutes per hour, while others require near-continuous operation. When operation is necessary, they are expected to perform efficiently and effectively, with a minimum of operator attention.

It is suggested that an owner/engineer contact operators of similarly sized screen installations to review the models operational history prior to making a final screen model selection. This is especially important in the case of unusual screen sizes or operating conditions.

NUMBER OF SCREENS

The number of Bar Screens required is usually a matter of the engineers preference. However, it is generally recommended that a minimum of two screens be installed. For smaller treatment plants, the second screen may be a simple, manually cleaned bar rack used during emergencies and installed in an overflow bypass channel.

Based on operating and maintenance considerations, two narrow Mechanically Cleaned Bar Screens are recommended over a single wide screen. When two or more screens are used, the channels and screening equipment should be designed so that one screen can be taken out of service without adversely affecting the operation of the remaining screen(s).

When possible, all screens at a single installation should be of the same model, manufacturer, and size. This will insure interchangeability of mechanical components and reduce the required spare parts inventory.

Future screening requirements must always be considered. If it is anticipated that additional Bar Screens will be required on future plant expansions, the size and location of the additional screens should be considered. It is usually cost effective to complete the civil work for future plant additions during initial plant construction.

BAR SPACING

The clear openings and width of the vertical bars has a direct bearing on screen efficiency. Screen efficiency is maximized when the greatest practical spacing between bars is coupled with the thinnest practical bars.The clear opening between bars on a Mechanically Cleaned Bar Screen usually ranges from 1/2" to 1-1/2". Bar widths generally range from 1/4" to 5/8".

The selection of the clear opening for a specific application is usually made on the basis of protecting downstream equipment or treatment processes. Most of the pumping equipment used in todays sewage treatment plants can tolerate those solids that pass a Bar Screen with 1" bar openings. The standard bar spacing on new treatment plants is generally 3/4".

For some plants, the increased "efficiency" of a finer opening may not justify the increased handling/disposal cost of the increased volume of screenings. Very close spacing of bars may also result in the removal of organic particles necessary to sustain downstream biological treatment processes.

VELOCITY

The size of the channel is determined by a review of the flow conditions. Generally, the channel should be designed to provide a minimum approach velocity of 1.5 feet per second to avoid sedimentation. Where large amounts of storm waters are to be handled, an approach velocity of approximately 3.0 feet per second should prevent grit from settling at the bottom of the Bar Screen.

The size of the actual screening area required for a Bar Screen is based on flow, bar rack efficiency, and the desired velocity through the bar rack. In most sewage applications, it is desirable to maintain a velocity through a clean bar rack at approximately 2.0 feet per second during periods of average daily flow. The screen should be designed so that the velocity through the bars does not exceed 3.0 feet per second during periods of maximum, or peak flow.

The velocity through the bars can be calculated according to the following formula:

$$A = (Q)\,(1.547) / (V)\,(E)$$

Where: Q = Flow (mgd)
V = Velocity (fps)
E = Bar Rack Efficiency (%)
and,
E = (Opening) / (Bar Size + Opening)

After determining the actual screening area required, the channel width and depth can be selected. For example, the screen area determined by the above formula may require 60 square feet of actual screen area. This could be configured in a bar rack measuring 6 feet wide and 10 feet deep, or 3 feet wide and 20 feet deep. It is recommended that sufficient freeboard (9" minimum) be added to the maximum water level to allow for flow surges.

A headloss develops across the screen as debris is collected on the face of the bar rack. This headloss should be relieved as the screen is operated and collected debris is removed. The theoretical headloss can be computed according to the following formula:

$$H = [(V^2 - v^2) / 2g]\,[1 / 0.7]$$

Where: H = Headloss
V = Velocity through bars
v = Approach velocity
g = Acceleration due to gravity (32.2 ft/sec/sec)

CHANNEL DESIGN

Mechanically Cleaned Bar Screens are usually installed side-by-side in concrete channels. Single screen installations will require a bypass channel with a manually cleaned bar rack for emergency operation. Channel designs should incorporate the use of stop logs or sluice gates to isolate each screen and facilitate dewatering for maintenance and inspection purposes.

The channel area upstream of the screen should be straight for several feet to insure that the flow is uniformly distributed on the face of the bar rack. The channel floor and sidewalls should be smooth and free of projections that might otherwise prevent uniform velocity distribution.

There are several methods of installing Bar Screens within a channel. Screens may be supported at the operating deck or channel invert. They may be inserted in recesses cast in the concrete channel walls, flush mounted to the walls and grouted in place, or installed within vertical steel guide slots. The arrangements available are dependent on the screen design selected, and screen manufacturers should be consulted for specific details.

Most screen channels are designed to be as shallow as practical to minimize the headloss through the treatment plant. The location of pumps, grit chambers, Parshall flumes, wet well, etc., downstream of the screen will affect the actual water depth in the channel. In some applications it may be necessary to install stop planks, acting as weirs on the downstream side of the screen, to increase the sewage depth.

Typical minimum and maximum channel widths for Mechanically Cleaned Bar Screens are 2'-0" and 14'-0", respectively.

SCREENINGS VOLUME

The volume of screenings removed by a Bar Screen varies greatly depending on the clear opening between bars and the character of the sewage. Normal variations for domestic sewage range from 0.5 to 10 cubic feet of screenings per million gallons of screened sewage. The screenings volume will obviously increase as the width of the openings are decreased.

For Bar Screens with openings of 1/2" to 1-1/2", the screenings volume can be expected to more than double with each 1/2" reduction in bar spacing. A bar screen having a 1/2" bar spacing will yield an average 7.5 cubic feet of debris per million gallons of screened sewage as compared to 1.25 cubic feet of debris for a screen with 1-1/2" openings.

Screenings usually contain approximately 80% moisture and weigh 40 to 60 pounds per cubic foot.

HANDLING AND DISPOSAL OF SCREENINGS

The screenings removed by Bar Screens in wastewater treatment plants are usually putrescible and offensive, and it is necessary that they are properly handled and disposed. Screenings may be collected in hoppers, cans or troughs that are an fabricated as an integral part of each screen unit. Multiple screen installations may be designed to share a common screenings conveyor belt, or trough to consolidate and convey screenings to a central point.

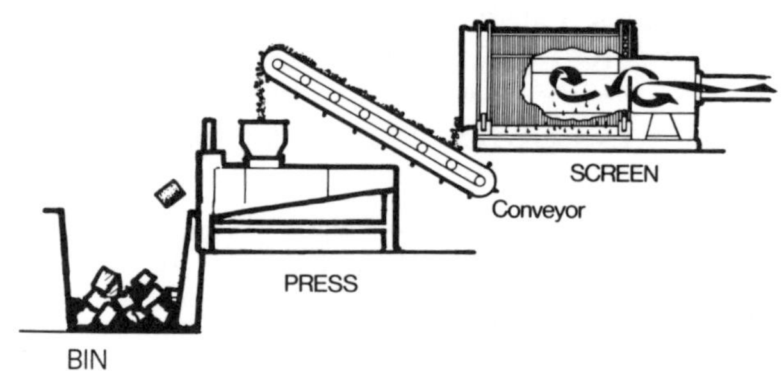

Screenings Handling and Disposal System
Courtesy of Eimco Process Equipment Co.

Screw-type and hydraulic screenings compactors can be used to further dewater, compress, and convey screenings. Depending on the type of screenings and the application, wet screenings can be pressed to 20% to 40% dry weight and reduced in volume by 75%. In addition, they may be automatically conveyed up to 30 feet horizontally or vertically.

Screenings Screw Compactor
Courtesy of John Meunier, Inc.

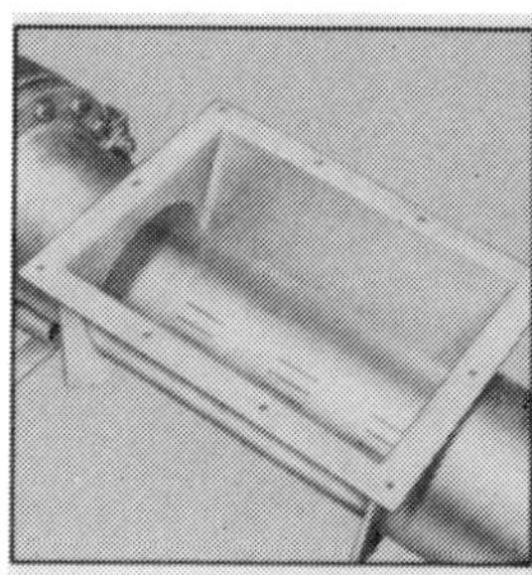

1 As screenings enter Hopper, free water drains through slots before pressing action occurs. A flange is provided on which chute can be fastened to direct screenings to the Hopper.

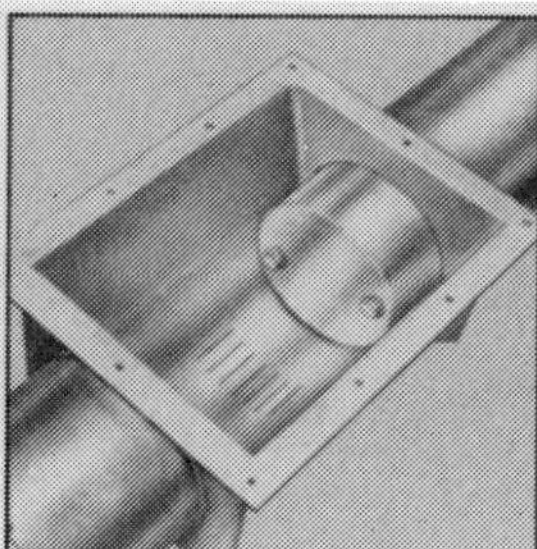

2 Hydraulic Ram smoothly advances and returns on a preset cycle of 30-90 seconds, depending on motor RPM and size of model.

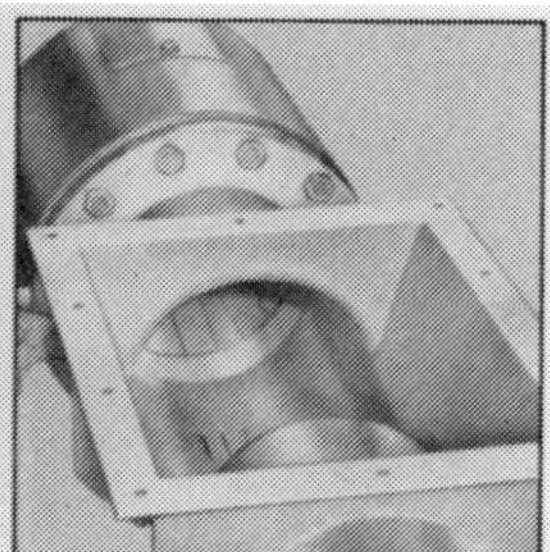

3 The wet screenings are dewatered and moved forward by Ram toward the Friction Cylinder.

Hydraulic Screenings Compactor
Courtesy of Hycor Corporation

Other methods commonly used to dispose of screenings, either alone or in combination include grinding, comminution, digestion, incineration, and burial.

SECTION 5

FINE SCREENS

The development of improved screening equipment is resulting in an increase in the application and use of screens with smaller, or "finer" clear openings. Fine Screens can be generally classified as screens having clear openings less than 1/2". They are used most frequently in wastewater treatment applications, including:

* *Solids removal to protect downstream equipment*
* *Primary/Secondary treatment, in lieu of clarifiers*
* *Solids recovery in industrial process streams*
* *Sludge screening*
* *Scum dewatering*
* *Centrate screening*

The use of a Fine Screen may have a significant impact on the sizing of downstream unit treatment processes. Some Fine Screens may reduce influent BOD levels by up to 35% or more. Because of this, the application of a Fine Screen should be carefully considered. It is recommended that the equipment design, sizing, and selection be discussed at length with potential equipment manufacturers. Pilot tests should be conducted on all "non-standard" liquid streams, and wherever Fine Screens are being used to replace existing unit treatment processes.

Like all other screens, the amount of debris removed by Fine Screens is related to the bar spacings. Fine Screens typically remove 6 to 32 cubic feet of debris per million gallons of sewage.

The reduced size of openings in Fine Screens may require that the screening surface be manually cleaned with brushes every 6 to 8 weeks. This may be especially true in applications encountering large amounts of grease in the screen feedwater.

CONTINUOUS ELEMENT FILTER SCREEN

A continuous element Filter Screen consists of individual screening elements mounted vertically on a series of horizontal shafts to form a endless cleaning grid, or belt, operating around headsprockets and a curved boot guide rail. The upstream face of the moving grid collects and retains debris, elevating it out of the flow and discharging into a suitable receptacle before beginning its descent.

Filter Screens are most commonly used on applications requiring 3/16" to 1/2" openings, although models are available with openings from approximately 1/8" to 1". They are offered for use in channel widths of 1 to 15 feet, and depths to 40 feet, and can be installed at inclinations of 5 to 30 degrees from vertical.

Discharge is accomplished on the rear side of the screen after the ascending rakes have rounded the headshaft.

There are two methods of discharging debris currently available. One design offers a combination drive/unloading headsprocket arrangement. The headshaft is equipped with a combination drive/unloading sprocket made of high impact resistant plastic. The periphery of the sprockets has flat blades that project through the cleaning elements from the inside the screen. The blades dislodge debris so that it can fall into the trash receptacle. A doctor blade assembly assists discharge by wiping the periphery of the sprocket as it projects through the elements.

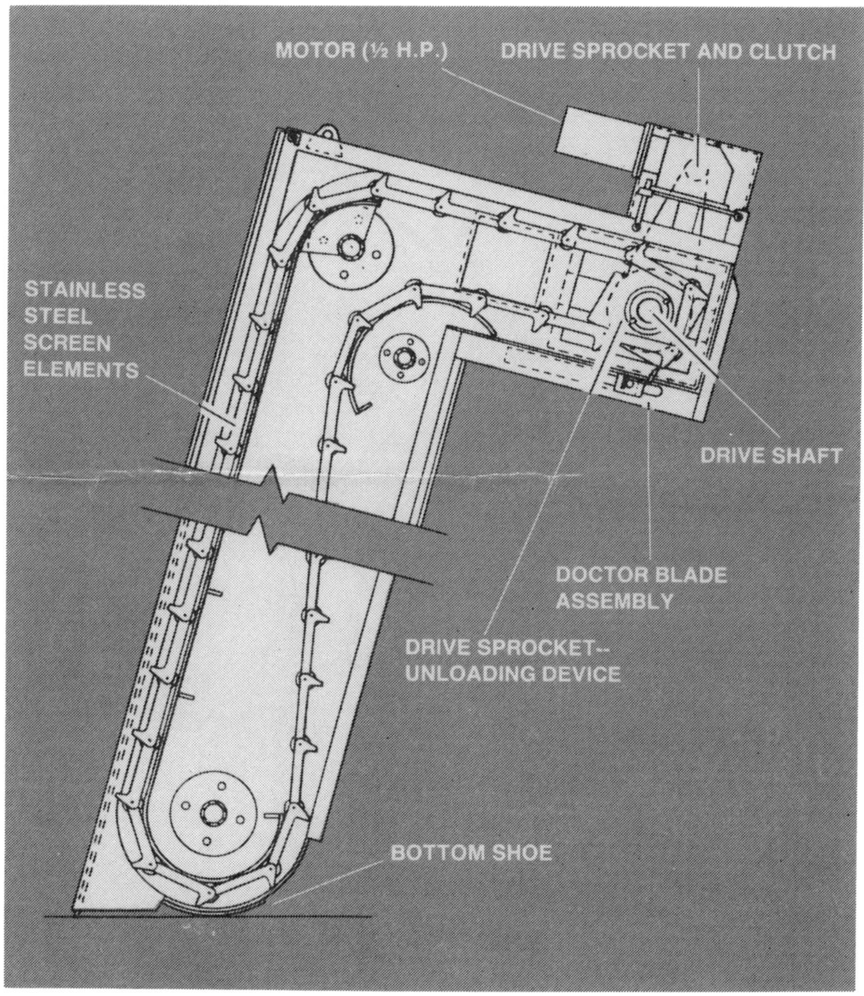

Continuous Element Filter Screen
Courtesy of Wiesemann Engineering, Inc.

A second cleaning method uses the rotation of the specially shaped cleaning elements to automatically discharge debris as the elements rotate around the headsprockets. The rotation causes a curved portion of the leading elements to move through the spaces of the trailing elements, ejecting debris. The action is then reversed as the belt rotates over a guide rail, causing the elements to wipe themselves clean. A rotating bristle brush is provided to assist with the discharge of stringy or sticky debris.

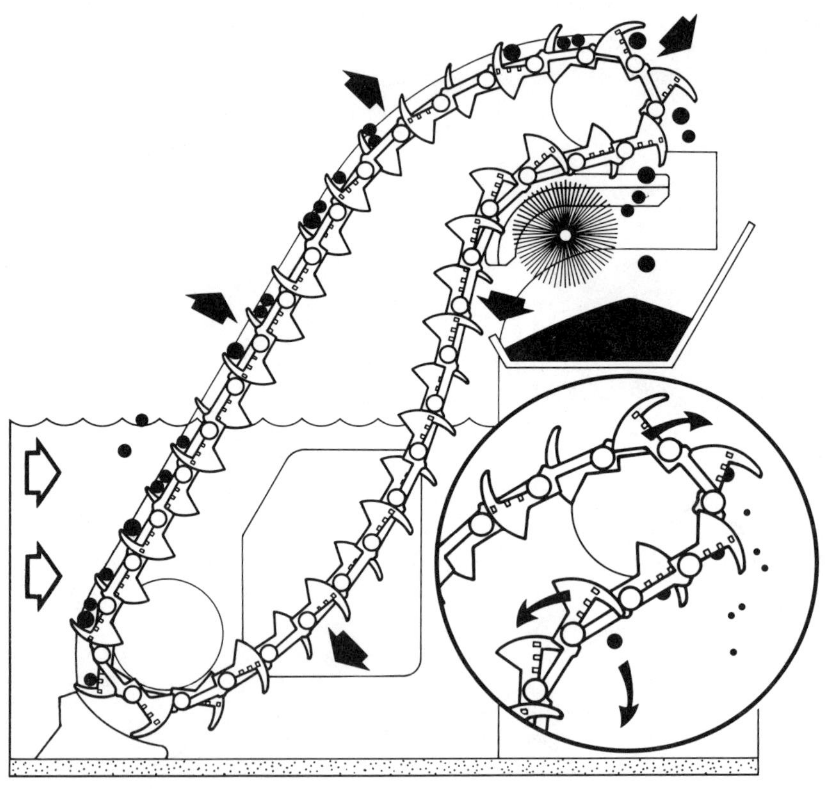

Continuous Element Filter Screen
Courtesy of Parkson Corporation

The cleaning elements may be constructed of stainless steel or an impact resistant plastic, and are shaped to accomplish both coarse and fine screening simultaneously. The elements are fitted to horizontal shafts equipped with guide rollers and closure plates on each end. The ascending guide rollers operate in u-shaped track guides fabricated as part of the screen frame. The closure plates seal the area between the outer cleaning elements and the edge of the screen frame.

The screen frame is fabricated from 3/16" thick steel, or stainless steel, and is supported at the operating floor level. The screens may be mounted on pillow block bearings to allow the entire screen to be pivoted out of the channel for maintenance or inspection. Side seals of neoprene prevent debris from bypassing the screen frame.

Power is introduced by an electric motor/gear reducer drive arrangement, and is transferred to the headshaft through a roller chain drive system. The drive motor is typically rated at 1/2 to 2 horsepower, and operates the screen at travel speed of 7 to 10 feet per minute.

ROTARY SCREENS

Rotary Screens utilize a wedgewire screen cylinder rotating on a horizontal axis to remove solids from a liquid stream. These self-cleaning screening devices are fabricated of stainless steel throughout, and are available with slot openings of 0.010" to 0.100" or more, and a standard wire width of 0.060".

Externally Fed Rotary Screen The externally fed Rotary Screen uses the outside periphery of the cylinder as its active screening surface. Influent enters the unit through a baffled headbox that distributes flow evenly along the length of the rotating screen cylinder. The liquid then gravity flows through the upper quadrant of the screening surface, into the center of the screen cylinder, and out through the bottom. Solids that are retained on the external surface of the screen are removed by a doctor blade as the cylinder revolves.

The doctor blade is held in contact with the screen surface by a spring loaded compression assembly, and can be furnished with an automatic cleaning mechanism. An optional internal spray backwash system can also be furnished.

The cylinder is supported by shafts on both ends, and operates in self-aligning pillow block bearings. A flexible seal located in the bottom of the headbox rides in contact with the rotating screen surface. End seals prevent debris from bypassing the screening surface.

Rotary Screens are available with 12", 24", and 36" diameter drums, and lengths from 1 to 10 feet. Hydraulic capacities range up to 10,000 gpm per unit. Screens can be located above or below grade, with a gravity or pumped feed to the headbox, and a gravity open channel or piped discharge. The 1/2 to 3 horsepower screen drive unit may be shaft mounted or direct coupled to the drive shaft, to operate the unit at 5 or 11 rpm. Smaller units may be air driven.

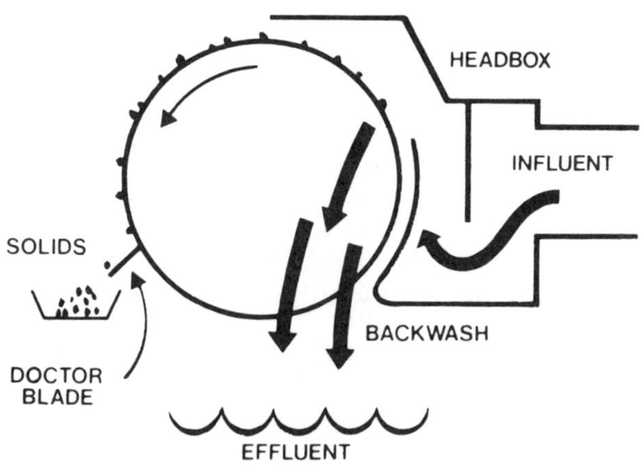

Flow Diagram of Externally Fed Rotary Screen
Courtesy of Hawker Siddeley Brackett

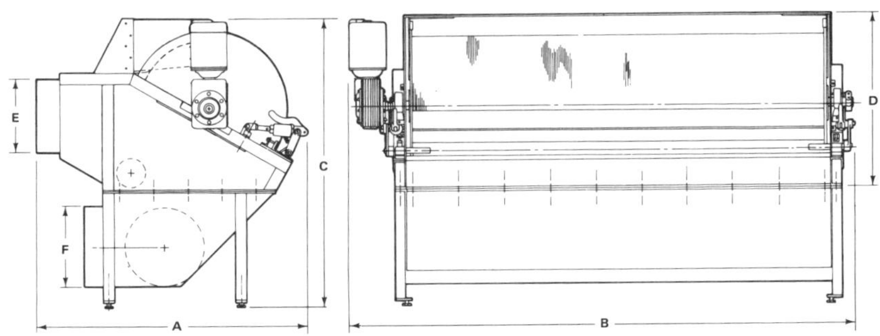

ALL DIMENSIONS IN INCHES										ALL WEIGHTS IN POUNDS			
Model No.	Motor Hp.	Screen Dia.	Screen Length	Unit Depth A	Unit Length B	Unit Height W/Base C	Unit Height W/O Base D	Influent O.D. E	Effluent O.D. F	Dry Weight W/Base	Dry Weight W/O Base	Operating Weight W/Base	Operating Weight W/O Base
RSA-2512	½	25	12	45.12	29.09	45.25	30.25	4.50	8.63	300	250	500	375
RSA-2524	½	25	24	45.12	41.59	45.25	30.25	6.63	10.75	600	525	1000	750
RSA-2548	¾	25	48	50.50	66.90	51.38	30.38	10.75	14.00	750	620	1550	1100
RSA-2572	¾	25	72	50.50	90.90	51.38	30.38	12.75	16.00	900	750	2100	1450
RSA-3672	1-2	36	72	70.75	92.50	67.38	43.75	16.00	20.00	1850	1650	5450	3250
RSA-36120	1½-3	36	120	70.75	142.44	66.75	43.75	20.00	18.00[2]	2940	2650	8450	5250

Externally Fed Rotary Screen

Courtesy of Hycor Corporation

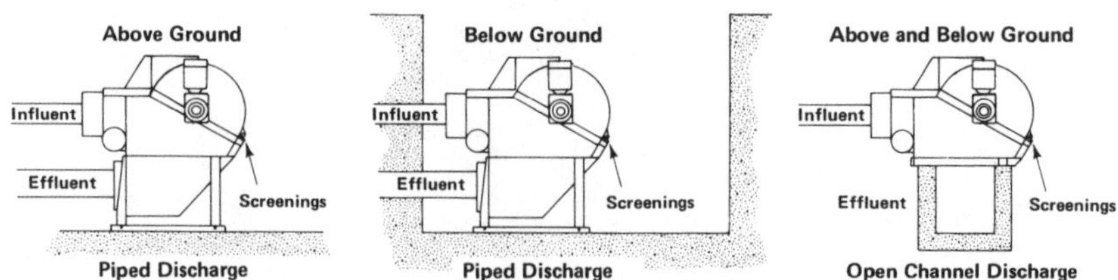

Layout Suggestions of Externally Fed Rotary Screens
Courtesy of Hycor Corporation

Internally Fed Rotary Screens The internally fed Rotary Screen utilizes the inside surface of a wedgewire screening cylinder as the active screening surface.

Internally Fed Rotary Screen
Courtesy of Hycor Corporation

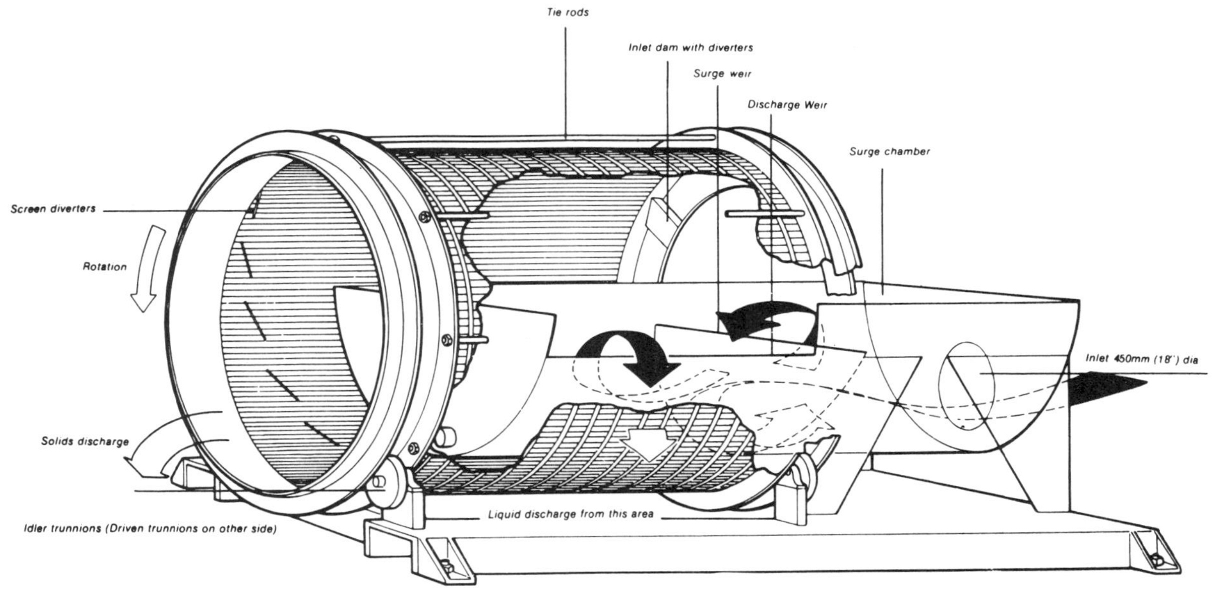

Internally Fed Rotary Screen
Courtesy of Eimco Process Equipment Co.

Influent enters the screen through a trough-type headbox in the center of the cylinder and overflows an adjustable discharge weir. As influent spills onto the sides of the rotating screen, the liquid passes through the slot openings and solids are retained on the inside face of the cylinder. Dewatering of the solids is enhanced by the tumbling action at the bottom of the rotating cylinder, and small diverter plates fixed to the inside of the drum move the solids axially to one end of the cylinder for discharge.

Internally fed screens are available in 24", 36", and 48" diameters, and cylinders up to 10 feet long, for flows up to 6000 gpm per unit. Slot openings usually range from 0.020" to 0.100". A 1/4 to 2 horsepower motor/reducer drive unit rotates the drum by means of a roller chain drive at approximately 6 rpm.

The cylinder is supported at each end by a pair of rollers, or trunnions. At least one manufacturer offers both an internal and external spray system, used on a continuous or intermittent basis, to clean the screen surface.

Internally Fed Rotary Screen Installation
Courtesy of Eimco Process Equipment Co.

In-channel Rotary Screen This screen is a variation of an Internally Fed Rotary Fine Screen that utilizes a cylindrical screening basket mounted in a channel at an angle of approximately 30 degrees. Flow enters the lower, open end of basket and passes through bars spaced to provide 0.2 to 0.6 inch openings. A toothed rake revolves around the interior of the screen and rakes debris to the top of the screen basket where it is removed by a cleaning comb. The debris falls into a trough equipped with a central screw conveyor/compactor and is elevated to the required discharge height.

There are at least two variations of this screen that are available with openings of 0.01 to 0.10 inches. One unit utilizes a rotating screen basket with an externally mounted debris spray, and another utilizes a semi-circular screen basket equippedwith a full diameter central screw to clean the bars and convey debris out of the channel.

Cylindrical Bar Screen with Screw-type Debris Elevator
Courtesy of Lakeside Equipment Company

DISC-TYPE FINE SCREEN

The Disc-Type Fine Screen consists of a flat disc covered with screening media that rotates about a horizontal axis, perpendicular to the flow. Influent enters the submerged portion of the disc and solids are retained by the screening media. The rotation of the disc lifts the solids above the water surface where they are removed by a bank of spray nozzles.

Disc Screens are available in sizes ranging from 6 to 16 feet in diameter with mesh openings as fine as 1/32". Applications are usually limited to those with flows less than 20,000 gpm.

Another type of Disc Screen uses a one or more pairs of slowly rotating, mesh-covered discs to accomplish removal of fibrous solids. As the discs rotate, solids form a "precoat" on the surface of the discs. The mat of solids rolls back by gravity as the discs rotate to self-clean the mesh. The debris exits the unit via a debris chute.

This screen is available with openings to 44 microns, and disc diameters to 60 inches.

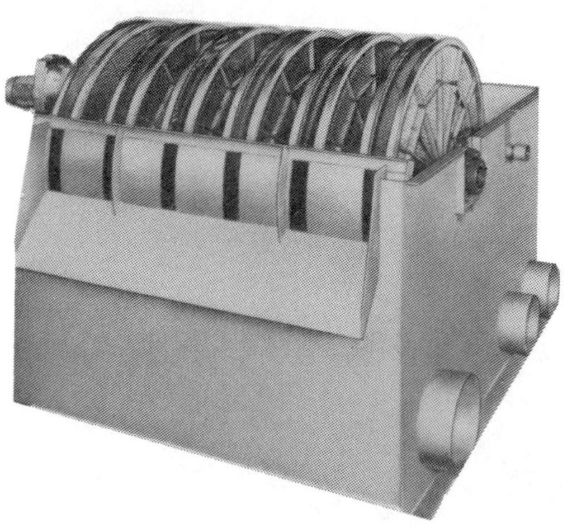

Multiple Disc Fine Screen
Courtesy of Hycor Corporation

BRUSH RAKED FINE SCREEN

Brush Raked Fine Screens operate similar to Arc Screens. The cleaning brush mechanism is mounted on a drive shaft that operates in pillow block bearings and slowly rotates in a complete 360 degree circle about a horizontal axis. The rotation of the cleaning brush cleans the radius of the stationary, perforated plate screen, and elevates solids to a discharge point at the top of the screen, where it is gravity discharged into a trough or container.

Horizontal shaft units may be up to 15 feet long with a screen radius of 3 feet.

This screen can also be furnished with a vertical shaft design for river intakes up to 16 feet deep with a screen radius of 18" to 5 feet. In these applications, the curved screening surface is recessed in the sides of a channel that is flush mounted with the river bank. As the rotating brushes clean the screen surface, debris is carried out into the river flow where natural currents can carry it away from the intake.

Perforated plate openings usually range from 1/8" to 3/4". Screens may be fitted with one or two sets of nylon or polypropylene cleaning brushes. Motor/reducer drive units may be direct coupled or shaft mounted.

STATIC SCREENS

Static Screens use a stationary, inclined screen deck that acts as a sieve to remove solids from liquids.

Influent enters the screen through a headbox located on the back of the unit. The headbox is appropriately sized and baffled to reduce turbulence and deliver flow to an overflow weir. Water flowing over the weir accelerates downward and passes over the inclined screen deck. As it cascades over the face of the screen, the liquid is stripped away from the solids and falls through the screen openings into a filtrate chamber. The oversized solids continue down the face of the screen and are gravity discharged from its lower edge.

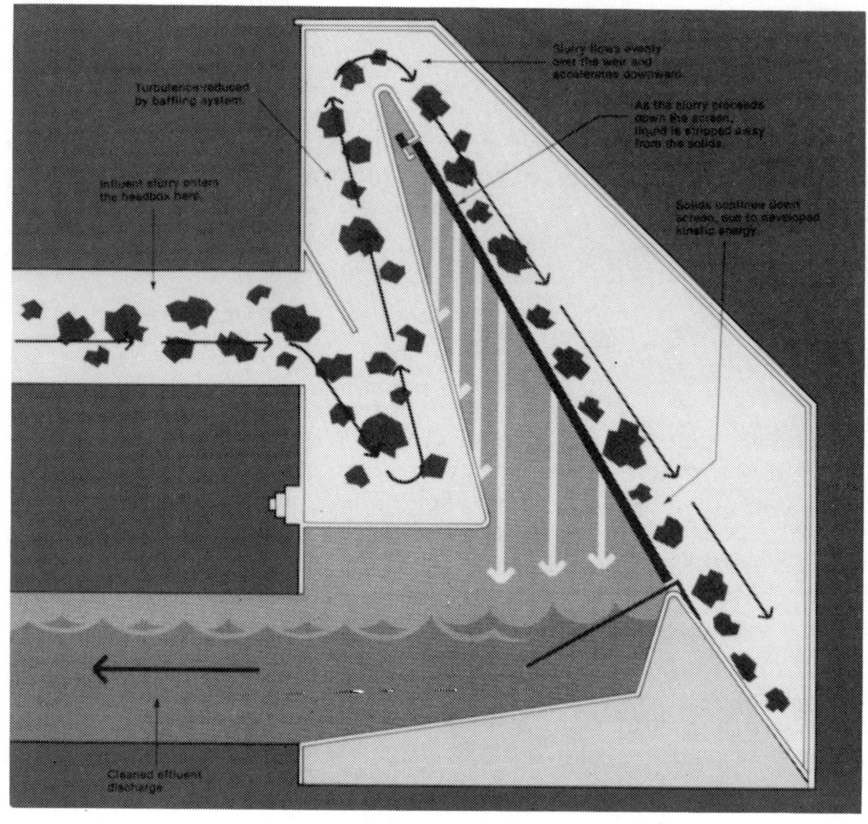

Static Screen
Courtesy of Hendrick Corporation

A combination of the bar design and deck inclination increases the hydraulic efficiency and dewatering ability of the screen.

The screen deck for these units is typically constructed of wedgewire bars with clear openings of 0.010" to 0.100". The slots are usually oriented horizontally. Screens are available in widths of 2 to 6 feet, with hydraulic capacities of over 1,000 gpm on the wide units.

Static Screen frames and screen decks are manufactured of stainless steel. Screen decks should be removeable, and most are adjustable over a range of approximately 45 to 60 degrees to suit site specific operating conditions. Several deck configurations are available. Designs include a straight screen, a curved screen, and a screen of three piece construction.

Other options available include a filtrate viewing port, lower drip lip, a pivoting screen deck, and a hinged flow distribution baffle.

SECTION 6

MICROSCREENING

Microscreening, or Microstraining, is a form of simple mechanical filtration employing a fine mesh fabric mounted on the periphery of a horizontally mounted, continuously, rotating drum.

Approximately 70 to 75 percent of the drum (66 percent of the screening area) is submerged under normal operation. Flow enters the drum interior through the open end and exits radially, through the screening mesh while solids are retained on the inner surface of the mesh. Captured solids are washed from the screening mesh into a backwash trough using a series of spray nozzles located at the top of the screen.

Microscreens are equipped with a filter mesh having openings that range from 1 to 200 microns, although most applications utilize openings within a 15 to 60 micron range. Screening is accomplished directly by capturing a solid on the surface of the screening media, or indirectly by capturing a solid on a "mat", or thin film of solids previously captured on the mesh surface.

The indirect capture of solids contributes to the efficiency of the screening process by reducing the effective size of the screening mesh.

The fine screening mesh used in Microscreen necessitates careful mechanical and process considerations in their application, design, and operation.

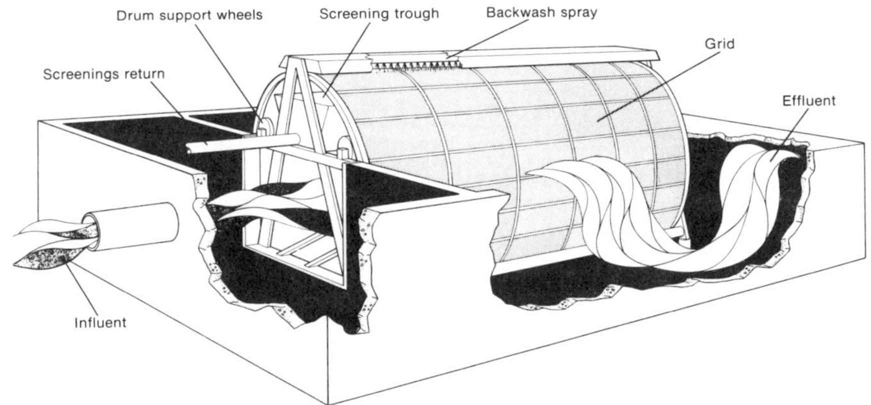

Microscreen Flow Diagram
Courtesy of Envirex, Inc.

Applications

Microscreens have been used in a variety of water and wastewater applications for more than 25 years. Increasing technology is renewing interest in the application of Microscreens to meet stricter discharge operating conditions.

Water treatment plants using a surface water supply may use Microscreens as an initial treatment step to remove a significant proportion of raw water suspended solids. This may include the removal of up to 75% of the algae and some free-swimming organisms, reducing chemical costs and increasing the performance of downstream treatment processes.

Some industrial process water applications, or direct filtration plants may use a Microscreen as the only other treatment step in their water treatment facilities.

Microscreens are most frequently used in wastewater treatment plants as a tertiary or final "polishing" process, although in several instances they have been used to replace secondary sedimentation. Effluent from aerated lagoons and oxidation ponds may also be polished with a Microscreen prior to discharge.

When used to polish secondary effluents, Microscreens may produce effluent qualities as low as 10 mg/l suspended solids and 10 mg/l BOD.

Microscreens can also be used to recover or reclaim materials from process waste streams in pulp and paper, meat packing, tanning and other industrial applications.

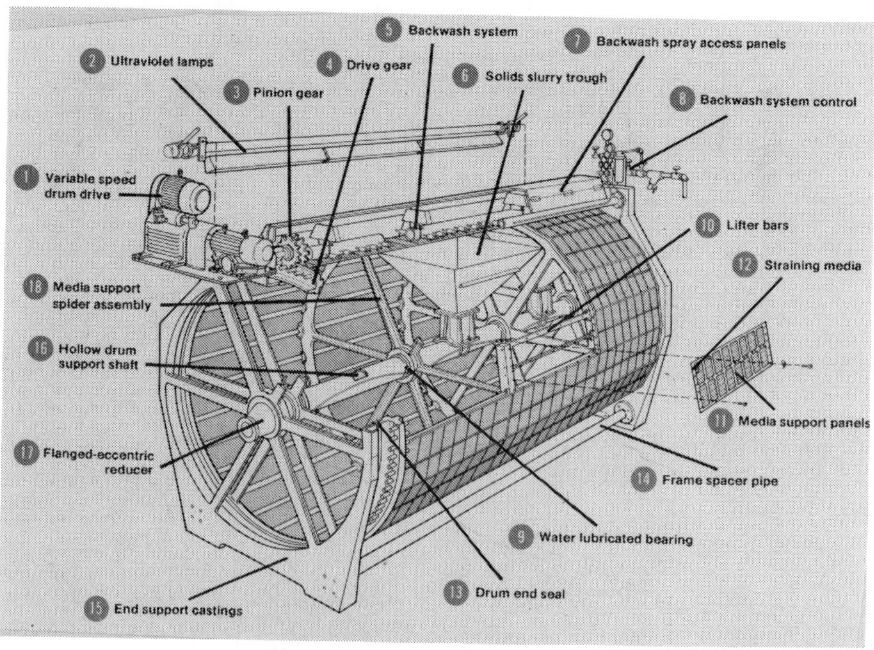

Microscreen Assembly

Courtesy of Lakeside Equipment Company

Mesh

The filter mesh (also referred to as media, fabric, or cloth) is the heart of the Microscreen unit. The size of the mesh openings will determine the overall size and process performance of the Microscreen unit.

A woven cloth of polyester wire is most commonly used as the screening media. The selection of mesh openings and wire diameters is application specific. Some common sizes of mesh openings are 1, 6, 15, 20, 25, 35, 50 & 60 microns. (It should be noted that individual manufacturers have limitations on the minimum mesh opening that they offer.) Although not as common as it once was, stainless steel wire mesh may also be used, but it is limited to a clear opening larger than 15 microns.

Mesh durability is important to the success of an installation. The combination of small openings and fine wire diameters result in a thin, flexible piece of mesh that requires intermediate support. Adequate support is usually provided by installing the mesh in small panels, or frames, each having a screening area that ranges from 2-1/2 to 10 square feet. These flat, or involute-shaped panels are then bolted to the Microscreen drum with easily removeable stainless steel fasteners.

Some Microscreen manufacturers offer proprietary panel designs that have been developed to overcome operational problems associated with fine mesh and facilitate panel installation, removal, and replacement. Most of these designs incorporate the use of a molded plastic mesh panel that is manufactured with an integral support grid. This square or hexagonal grid supports the mesh at 1 to 3 inch centers.

Drum and Support Frame

Microscreens are available in diameters that range from a 6 to 12 feet, and lengths ranging from 4 to 18 feet. Manufacturers also offer smaller, "packaged" Microscreens in steel tanks for low flows.

Design and construction of the drum assembly and support frame varies among equipment manufacturers. The basic mesh support structure is a cylindrically-shaped, suitably stiffened rigid frame fabricated of carbon or stainless steel. End supports/plates may be fabricated or cast, of carbon steel, stainless steel, or aluminum.

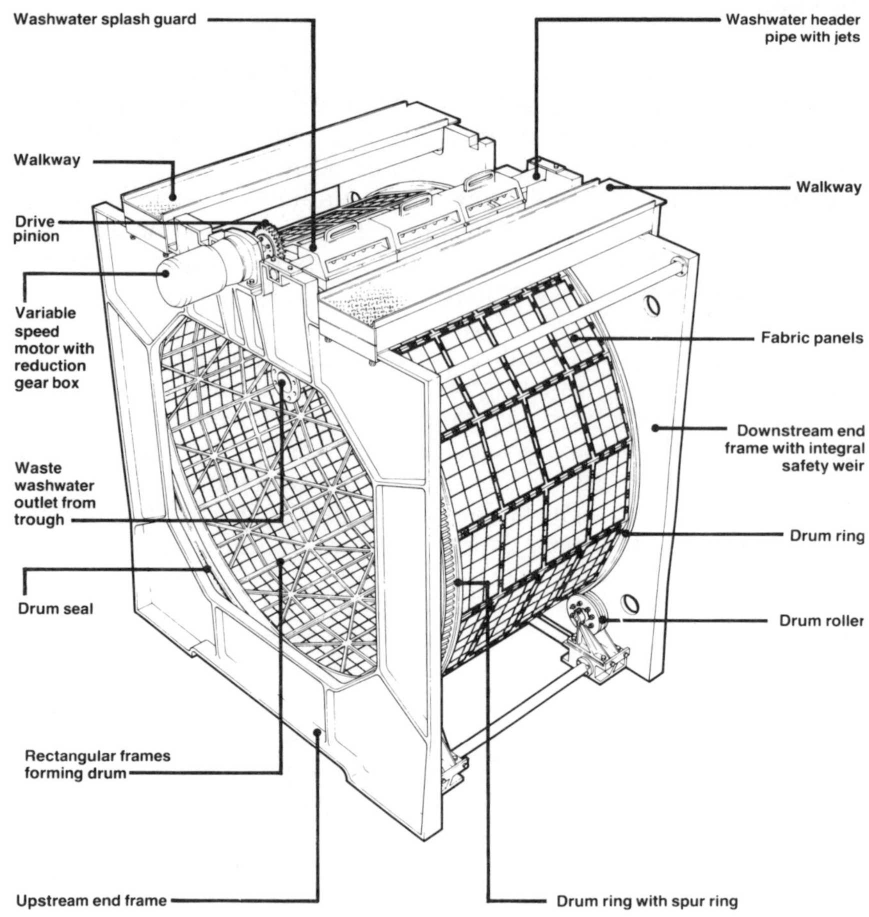

Microscreen Assembly
Courtesy of Mather & Platt, Ltd.

There are several methods used to support the drum assembly. One method utilizes a pair of nylon rollers to support the upper inside surface of the inlet side of the drum. The closed end of the drum is supported by a stub shaft operating in water lubricated thrust bearing. Another manufacturer assembles the ends of the drum structure to cast bronze end rings that are supported by tire-like support rollers. Other designs include the use of a central shaft to support the drum frame, and a support arrangement using a large diameter journal bearing located in the end frame to support a bearing flange formed as part of the drum end plate.

Unscreened water is prevented from passing around the both ends of the drum by a flexible seal. At least one manufacturer offers a seal arrangement that can compensate for wear and can automatically increase sealing pressure in response to an increase in headloss.

Backwash Spray System

Debris is removed from the drum by means of a pressurized water spray. A spray pipe header, equipped with replaceable nozzles, is located above the drum, directing a water spray through the mesh and into a backwash trough located inside the drum.

The backwash water is usually obtained from the screened Microscreen effluent. Backwash water requirements may range from 2 to 5 percent of the screens total throughput, at a pressure of 15 to 70 psi.

Spray systems may be designed and operated to suit a particular, or varying operating condition. Options include single or dual spray pipes, for constant or intermittent operation, at a constant or variable pressure.

Periodic spray nozzle purging may be required. It is recommended that self cleaning spray nozzles be utilized to reduce maintenance and increase process reliability.

The spray system is covered with a corrosion resistant spray hood to prevent overspray. Most manufacturers offer hoods equipped with clear plastic inspection windows for visual observation of the operating spray system. A walkway or bridge is recommended to provide access for maintenance and inspection of the spray system.

The backwash trough, located at the inside top of the drum, should be designed with steeply sloping sides to prevent debris accumulations.

Microscreen Assembly
Courtesy of Mather & Platt, Ltd.

Supplemental Mesh Cleaning

Over a period of time the Microscreen mesh may become clogged with oil, grease, or biological growths. Supplemental mesh cleaning will be required to reduce this clogging. Depending on the application, manufacturers suggest cleaning the mesh at intervals that range from every three to eight weeks of operation.

Several cleaning methods are available based on the type of fouling that has occurred. When selecting a supplemental cleaning system, its compatibility with the mesh material should be considered. These cleaning methods include the use of mild chlorine solutions, ultra-violet irradiation, steam, and high pressure hot or cold water sprays.

Mesh Inspection
Courtesy of Mather & Platt, Ltd.

Each cleaning system has advantages and disadvantages. The use of chlorine will require that the screens be removed from service for up to eight hours or more. Cholrine also has a deleterious on stainless steel media. Ultra-violet irradiation may require frequent lamp replacement and is effective only on the exposed surfaces of the biological growths. UV irradiation is also harmful to polyester mesh. High pressure water or steam cleaning is also expensive and time consuming.

Drive & Controls

Microscreens are rotated at a peripheral drum speed that ranges between 15 and 150 fpm. Rack and pinion, drive chain and sprocket, and v-belt and sheave arrangements may be used to transfer power from an electric motor-speed reducer drive unit to the drum.

A rack and pinion drive gear system consists of a nylon or bronze pinion gear keyed to the output shaft of the speed reducer. The pinion engages a larger gear ring fixed to the periphery of the drum's drive end. Designs utilizing segmented cut tooth gears, and pin rack type gear rings are available.

Power can also be transferred to the drum by means of a drive chain/v-belt operating over a drive and driven sprocket/sheave. The drive sprocket is mounted on the output shaft of the reducer, and the driven sprocket is mounted to a stub, or central shaft fixed to the drum.

The drum's rotational speed should be adjustable, and may be automated to accommodate sudden variations in raw water quality and quantity. Variable speed may be accomplished using a DC drive motor, variable frequency AC drive motor, or transmission.

In most applications it is preferred that a constant headloss across the screen be maintained. To accomplish this, the screen control system should be capable of varying the rotational speed of the drum as the differential headloss across the screen increases or decreases. Rotational speed can be controlled using the output signal from a pneumatic or capacitance-type differential level controller.

Additional controls can be furnished to regulate backwash operation and provide other specific process requirements.

Sizing & Process Considerations

The mechanical process of Microscreening is fairly simple. However, it is very hard to predict the effectiveness of the process for a specific application. Microscreens are usually sized on the basis of hydraulic loading of the submerged screen area (i.e. gpm/sq ft) at a given headloss. Determination of the loading for a specific application is made based on a thorough review of the application, data from similar installations, and/or pilot and bench scale testing.

Typical hydraulic loadings range from 5 to 20 gpm/sq. ft. of submerged area at 12" of headloss. However, units with openings of 1 to 7 microns may be rated at 1.5 gpm/sq. ft. at up to 24" of headloss. Industrial, and larger mesh size applications may utilize loadings of 50 gpm/sq.ft. or more.

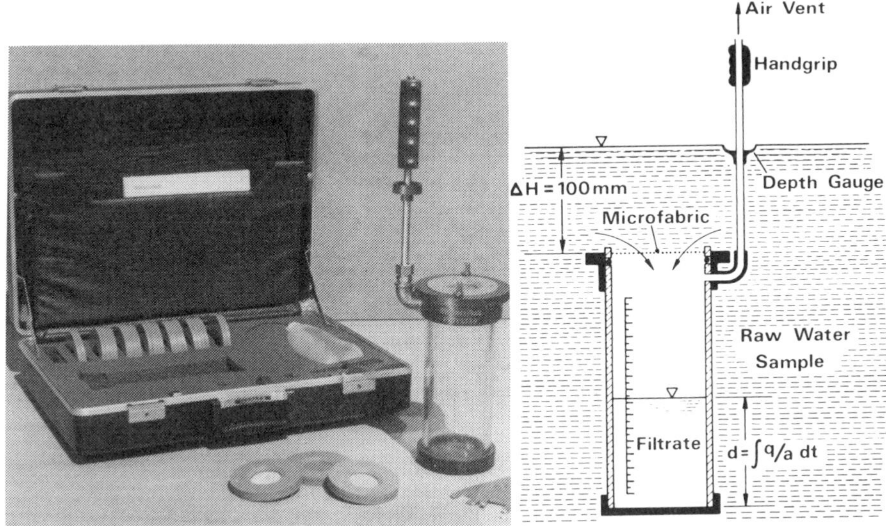

Microscreen Field Test Kit
Courtesy of Mather & Platt, Ltd.

Most manufacturers have developed filterability test kits that can be used to determine the effectiveness of a Microscreen for individual applications. These field test kits can be used as guides in the selection of mesh size and hydraulic loading, and can reasonably predict effluent quality.

All pilot testing should be done during periods of maximum solids loading, and simulate actual operation (i.e. location, temperature, pretreatment, etc.) as closely as possible.

SECTION 7

SCREEN CONTROL SYSTEMS

The purpose of efficient screen control is to hold the differential headloss and velocity across the screen at or near the optimum "clean screen" levels. Although some screens may be able to be operated manually, most installations require some type of automated controls to insure reliability and economy of operation.

There are few limits to the versatility that can be designed into a screen control system. This chapter will briefly describe typical control systems offered by screen manufacturers.

HAND-OFF-AUTO

A simple screen control system may consist of a Hand-Off-Automatic (HOA) switch. In the "Hand", or manual position, the screen will continue to operate until turned to the "Off" position. When the screen is operated manually, the optional automatic control devices such as the time clock, limit switch, float switch or differential level indicator have no effect.

In the "Automatic" position, the screen operation will be initiated by one of the optional control devices mentioned above.

FLOAT SWITCH

Many municipal bar screen installations utilize float type switches to start and stop screen operation.

Additional float switches can be located at various elevations to operate annunciators, pumps, or change the travel speed of screens equipped with multi-speed drive units.

DIFFERENTIAL LEVEL CONTROLLER

The purpose of a differential controller is to monitor the water level on the upstream and downstream sides of a screen and automatically perform a series of predetermined functions based on the differential water level.

Differential controllers should be equipped with a minimum of two pressure switches or operational setpoints to indicate a "low" and "high" differential levels.

Most differential controllers are designed to activate the screen drive motor at a preset level of approximately 6 inches. If the headloss is not relieved, and continues to rise to a second preset level after the screen has been started, an alarm is sounded to warn of a potentially damaging headloss condition. Screens with two speed drive motors may be switched to "high speed" at this higher differential level.

The controller should be designed to allow the screen to complete its operating cycle after the headloss drops below the "low" level. Traveling Water Screens and multi-rake Bar Screens should be furnished with an adjustable timer that will insure that the screen continues to operate for 1-1/3 revolutions after the headloss is relieved to prevent excess debris from drying out on the trays or rakes.

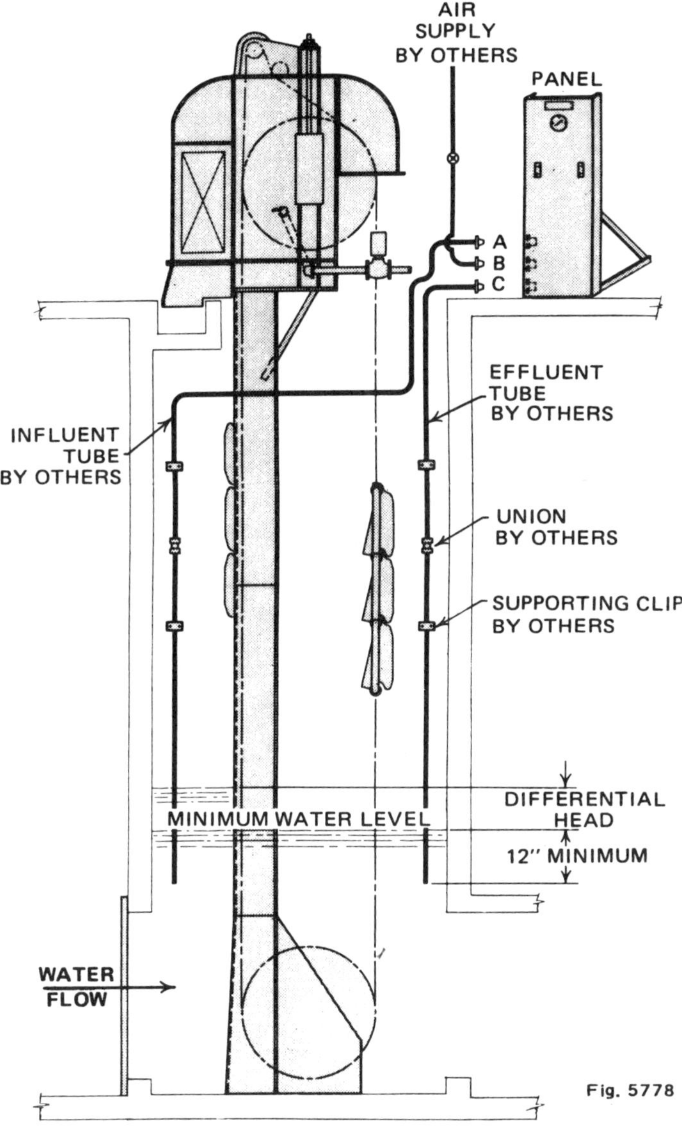

Differential Water Level Control System
Courtesy of FMC Corporation

Controllers can be designed with additional level set points to shut down pumps at very high headloss levels, or initiate operation of other equipment or alarms. Control systems for Traveling Water Screens should be designed to start the screen wash pumps, and/or open screen wash water lines prior to starting the screen. It is also recommended that the system be equipped with a pressure sensing switch to insure that adequate line pressure is developed before the screen is rotated.

Pneumatic Differential Controller This controller monitors headloss using two equally submerged vertical standpipes, or dip tubes, located on the upstream and downstream sides of the screen. A constant flow of air is delivered to the standpipes by means of a common pressure regulator to keep them purged of water. The difference in air pressure required to maintain the pipes in this condition causes a back pressure on both sides of a differential pressure switch. As long as the water level remains the same on both sides of the screen, the back pressure remains equal and no differential is indicated.

As debris accumulates on the face of the screen, the downstream water level begins to fall. The change in water level is indicated as a difference in back pressure by the pressure switch. When the differential water level increases to a setpoint level, the pressure switch is closed and the screen activated.

To prevent the submerged ends of the standpipe from becoming clogged with marine growth or other debris, air pressure should be maintained. It is also suggested that the tubing be installed as vertical as possible to allow cleaning with a push rod.

Ultrasonic Differential Controller This systems uses two ultrasonic transducers to measure the water level on the upstream and downstream sides of the screen. Both transducers transmit simultaneously and a differential analog meter measures the time differential between the echoes to determine the water level differential.

It is recommended that the transducers be mounted in stilling wells to minimize activation of the control unit by wave action.

Head Pressure Sensing System This system utilizes pressure transducers that react to changes in water elevation by measuring the head pressure on the upstream and downstream side of the screen. The continuous level readings are sent to a differential control unit that controls the screens operation as a function of the headloss differential.

SECTION 8

SELECTION OF MATERIALS

The selection of the proper materials of construction can be one of the most important aspects of screening equipment design. Many screen failures are attributable to poor material selections of various screen components. With an increasing variety of materials being offered, including over 60 types of stainless steel, the selection of suitable materials of construction can be a challenging assignment.

The selection of a suitable material should involve a review of five important design criteria. These criteria include the fabrication requirements, and the materials total cost, corrosion resistance, mechanical properties, and product availability.

This section briefly describes the more common materials and alloys used in the manufacture of water and wastewater screening equipment.

STAINLESS STEELS

This is a general term that is usually used to describe steels that usually contain a minimum of 12% chromium as the principle alloying element. Greater corrosion resistance can be provided by increasing the chromium content up to 27%.

Type 302 and 304 Familiarly known as "18-8 stainless steels", these alloys are popular austenitic stainless steels containing 18% chrome and 8% nickel. They are nonmagnetic steels which cannot be hardened by heat treatment, and instead, must be cold worked to obtain higher tensile strengths. They have excellent corrosion resistance, and can be readily fabricated by all methods usually employed with carbon steels. Type 304 contains a maximum 0.08% carbon to make slightly more corrosion resistant than Type 302.

Type 304L This is similar to the other 18-8 alloys except for a lower carbon content of 0.03% maximum, reducing the possibility of carbide precipitation. The corrosion resistance is therefore unaffected by normal welding and stress relieving applications. Type 304L is usually used if extensive welding is required in fabrication.

Type 316 This is a modified version of Type 304 containing 2% to 3% molybdenum. The addition of molybdenum increases the corrosion resistance of this alloy and makes it less susceptible to pitting and pin hole corrosion by sea water, brine, and chlorides.

Type 316L This is similar to Type 316 except for a lower carbon content of 0.03% maximum, reducing the possibility of carbide precipitation. Type 316L is the preferred grade of Type 316 stainless steel if extensive welding is required in fabrication.

Type 410 This is a basic 400 series stainless steel alloy containing 12% chrome, that is magnetic and can be readily heat treated to provide a wide range of mechanical properties. It is fairly resistant to mild forms of corrosion.

Type 416 This alloy is closely related to Type 410 with similar mechanical properties, and slightly lower corrosion resistance. One or more elements, usually phosphorous and sulphur, are added to improve machinability.

Type 420 This is a 12% chromium steel with a nominal 0.30% carbon content that results in better responsiveness to heat treatment than Type 410. It must be hardened and ground to obtain maximum corrosion resistance.

Type 440 A hardenable 17% chromium steel that offers better resistance to wear than Type 420. Available in several grades with varying carbon content to provide hardness levels to Rc 60/62.

Type 17-4 PH This martensitic alloy contains 17% chromium and 4% nickel and is usually superior in corrosion resistance to the 400 series stainless steels. 0.04% copper is added to promote precipitation hardening (PH) capabilities.

Alloy 20 A grade of Stainless Steel that is used for its improved corrosion resistance, especially corrosion by sulfuric acid. It contains approximately 20% chromium and 33% nickel in addition to small amounts of colombium.

OTHER MATERIALS USED IN SCREEN MANUFACTURE

Abrasion Resistant (AR) Steel This is a medium-carbon, high-manganese steel with excellent abrasion resistant properties and better workability than carbon steel of the same hardness.

Aluminum Aluminum is a light weight metal that has excellent corrosion resistance and electrical conductivity. It weighs approximately 28% as much as carbon steel and possesses a high strength-to-weight ratio.

Aluminum Bronze Common name for a series of copper alloys containing 5 to 12% aluminum. Aluminum bronze alloys containing approximately 10% aluminum are popular because of their good mechanical properties.

Babbitt Commercial name of a group of alloys used primarily as bearing materials. These materials may be tin or lead based, alloyed with antimony, arsenic or copper.

Brass An alloy of copper and zinc containing up to 40% zinc with small amounts of other metals added to obtain increased strength, hardness, or corrosion resistance.

Bronze The term bronze is generally applied to any copper alloy that has a metal other than zinc or nickel as the principal alloying element. Some brasses are called bronzes because of their color, or because they contain tin.

Buna N A nitrile elastomer, or NBR rubber known for its outstanding resistance to oil and fuel.

Carbon Steel A steel whose major properties depend on its 0.1 to 2% carbon content without substantial amounts of other alloying elements.

Cartridge Brass A commonly used copper alloy with a content of approximately 70% copper and 30% zinc.

Cast Iron The metallic product that is obtained by reducing iron ore with carbon at a temperature that is high enough to render the metal fluid and cast it in a mold. Also called pig iron, it is generally hard and brittle, and is neither malleable or ductile.

Commercial Bronze A copper alloy that contains approximately 90% copper and 10% zinc. Although commonly called bronze, it is actually a brass.

Copper Copper is a widely used metallic element. Its corrosion resistance, strength, fatigue resistance, formability and electrical conductivity make it, and its alloys well suited for a variety of applications.

Cupro-nickel A copper alloy containing 10% to 30% nickel that has excellent corrosion and stress corrosion cracking resistance.

Gray iron Cast metal widely used in sprockets and machinery parts that consists essentially of iron and carbon, with a relatively large portion of its carbon in the form of graphite.

Malleable Iron A ferrous material that consists primarily of temper carbon and soft iron, and includes small amounts of silicon, manganese, phosphorus, and sulfur. It offers strength, ductility, and machinability at a relatively low cost.

Monel An alloy of nickel and approximately 30% copper, with good mechanical and corrosion resistant properties. K monel, a precipitation hardened form of monel, has high resistance to impact and vibrational stresses.

Neoprene A synthetic rubber, or elastomer, that is chemically, physically, and structurally similar to rubber.

Polyethylene, UHMW Ultrahigh-molecular-weight polyethylene. Self lubricating plastic with low coefficient of friction, and excellent abrasion and impact resistance.

Polyurethane A copolymer characterized by good tensile strength, tear strength, and abrasion resistance. Polyurethane elastomers possess the rigidity of plastics and the resilience of rubber. They can be formulated for injection molding or casting.

Red Brass A copper alloy with good cold working properties and superior corrosion resistance that contains approximately 85% copper and 15% zinc.

Silicon Bronze A bronze that contains small amounts of silicon and manganese. Silicon bronze has greater strength and weldability than copper with comparable corrosion resistance.

Stellite A hard, wear- and corrosion-resistant family of nonferrous alloys containing cobalt (20-65%), chromium (11-32%), and tungsten (2-5%).

Tool Steel Steel containing a high carbon and alloy content characterized by high hardness and resistance to abrasion.

White Iron Iron containing 2 to 6% carbon, present as iron carbides. White Iron cannot be worked, but it has good casting properties and has applications where a cheap material with hardness and wear resistance is required.

Wrought Iron Wrought iron is one of the purest grades of commercial iron, containing approximately 0.1% of carbon and very small percentages of other non-metals. It is malleable, ductile, and flexible, and can be forged and welded.

EFFECTS OF ALLOYING ELEMENTS

Aluminum (Al) Aluminum is a common deoxidizer used in the manufacture of steel to reduce the oxygen content and prevent a reaction between carbon and oxygen during solidification. It is also used to control grain size.

Boron (B) Boron can be added to steel in amounts of 0.0005 to 0.003% to improve hardenability. When used in combination with other alloying elements, boron can help increase the depth of hardening during quenching.

Carbon (C) Steel owes it properties chiefly to various amounts of carbon. As the amount of carbon increases up to 0.8 or 0.9%, the metal becomes harder and possesses greater tensile strength, and becomes more responsive to heat treatment.

Chromium (Cr) Chromium is added to alloy steel in amounts of up to 1.5% to increase hardenability. If added in amounts greater than 4%, chromium confers an ability to resist corrosion. Stainless steels generally contain 12 to 30% chromium.

Cobalt (Co) Cobalt increases the strength and hardness of steels and allows the use of higher quenching temperatures. It is also used to intensify the effects of other major elements in more complex alloys.

Columbium (Nb) The use of columbium in 18-8 stainless reduces harmful carbide precipitation and resultant inter-granular corrosion. Welding electrodes containing columbian are used in welding both titanium and columbium bearing stainless steels since titanium would be lost in the weld arc, whereas columbium is carried over into the weld deposits.

Copper (Cu) Copper is normally added in amounts of 0.15 to 0.25% to improve resistance to atmospheric corrosion and to increase tensile and yield strengths with only a slight loss in ductility. Higher strength properties can be obtained by precipitation hardening copper-bearing steel.

Iron (Fe) Iron is the chief element of steel. Commercial iron usually contains other elements present in varying quantities which produce the required mechanical properties. Iron lacks strength, is very ductile and soft and does not respond to heat treatment to any appreciable degree. It can be hardened somewhat by cold working, but not nearly as much as even a plain low carbon steel.

Lead (Pb) Lead is used to improve the machinability of steel. When the lead is finely divided and uniformly distributed it has no known affect on the mechanical properties of the steel in the strength levels most commonly specified. It is usually added in amounts from .015% to 0.35%.

Manganese (Mn) Manganese is an important alloying element present in all steel. It functions as a deoxidizer and degasifier, and is also used to impart strength and responsiveness to heat treatment. Manganese is usually present in quantities from 0.5% to 2%, although some specialty steels may contain as much as 10% to 15% manganese.

Molybdenum (Mo) Molybdenum increases strength, hardness, hardenability and toughness. Molybdenum improves machinability and resistance to corrosion, and is an important means of assuring high creep strength. It It is generally used in comparatively small quantities ranging from 0.10 to 0.40%, and it intensifies the effects of other alloying elements.

Nickel (Ni) Nickel increases strength and toughness. Steels containing nickel usually have more impact resistance at low temperatures. Certain stainless steels employ nickel up to about 20%.

Phosphorus (P) Some phosphorus is present in all steel. It increases yield strength, hardness, and improves machinability.

Silicon (Si) Silicon is one of the most common deoxidizers and degasifiers used during steel manufacture. It also may be present in various quantities up to 1% in the finished steel and has a beneficial effect on certain properties such as tensile and yield strength. It is also used in special steels to improve the hardenability and forgeability.

Sulphur (S) Sulphur increases machinability in free cutting steels. Sulphur is detrimental to the hot forming properties.

Titanium (Ti) Titanium is a stabilizing element added to 18-8 stainless steels to make them immune to harmful carbide precipitation. It is sometimes added to low carbon sheets to make them more suitable to porcelain enameling.

Tungsten (W) Tungsten is used as an alloying element in tool steel and tends to produce a fine, dense grain and keen cutting edge when used in relatively small quantities. When used in quantities of 17 to 20%, and in combination with other alloys, it produces a high speed steel which retains its hardness at the high temperatures developed in high speed cutting. It is usually used in combination with chrome or other alloying elements.

Vanadium (V) Vanadium, usually in quantities from 0.15 to 0.20%, retards grain growth, permitting higher quench temperatures. It also intensifies the effects of other alloying elements.

SECTION 9

CORROSION PROTECTION

The design, construction and operation of a water or wastewater screening facility represents a major capital expenditure. Corrosion related screening problems can significantly increase plant operating and maintenance costs and may even result in plant shutdowns.

Screening equipment is especially susceptible to corrosion. Some of the factors which contribute to the high corrosion rates experienced include:

* The screen mechanism usually is constructed with a variety of metals producing significant dissimilar metals corrosion.

* Spray washing of debris collected on some screens removes the corrosion products which would otherwise tend to inhibit or reduce the corrosion rate.

* The water which passes through the screens is often of low resistivity (salt, brackish or contaminated fresh water).

Other factors that affect the rate of corrosion include changes in water chemistry and velocity, exposure to microbiological organisms, and varying dissolved oxygen levels.

There are several methods that can be used to combat corrosion. Some common methods used to prevent corrosion include the careful selection of materials, the use of protective coatings, and cathodic protection.

GALVANIC CORROSION

Galvanic corrosion is the natural deterioration of a material as a result of its interaction with its environment.

When two dissimilar metals or alloys are present in an electrolyte, a potential corrosion problem exists. If the two dissimilar metals are connected electrically, a current will be generated and one of the metals, the less noble metal, will corrode. Current from the corroding metal will flow into the electrolyte, over to the non-corroding metal, and then back through the connection between the two metals. The corroding metal is the one where the current leaves to enter the electrolyte and is known as the anode; the metal which receives current is known as the cathode.

Metals close to each other in the electrolytes galvanic series table are less severely attacked than metals separated by several steps in the table.

PROTECTIVE COATINGS

Painting Painting is the most common method of minimizing corrosion on carbon steel screening equipment. Coal tar epoxy and epoxy-polyamide paints are the most frequently used types of paints for submerged surfaces on screening equipment.

These paints are chemical and corrosion resistant coatings that form a durable, heavy film barrier against physical damage caused by direct impact, abrasion and flexing.

Surfaces to be coated will usually require sandblasting prior to being painted with two coats of paint, each having a dry film thickness of 6 to 10 mils. Although many coal tar epoxies and epoxy-polyamides are paints are self-priming, special primers are sometimes used.

Galvanizing Galvanizing is another common method of protecting carbon steel surfaces from corrosion. Galvanizing refers to the process of coating steel with a layer of zinc to prevent corrosion.

Zinc may be applied to an article by immersing it in a bath of molten zinc (hot-dipping), by electrodeposition (electrogalvanizing), or by metal spraying (metallizing).

CATHODIC PROTECTION

courtesy of Kevin C. Garrity, P.E., Ron Bianchetti, P.E. and Harco Technologies Corporation

Cathodic protection is an electrical method of preventing corrosion. It operates by passing direct current continuously from electrodes which are installed in the electrolyte, to the structure to be protected. Corrosion is arrested when the current is of sufficient magnitude and is properly distributed.

There are basically two methods of applying cathodic protection although there are numerous variations of these methods. One of these methods, usually called an impressed current system, uses anodes which are energized by an external DC power source. In this type of cathodic protection system, anodes are installed in the electrolyte and are connected to the positive terminal of the DC power source; the structure which is to be protected is connected to the negative terminal of that source.

The second method of protection makes use of galvanic anodes which have a natural difference of potential with respect to the structure to be protected. These anodes are made of a material, such as magnesium or zinc, which is anodic with respect to the protected structure and these anodes are connected directly to that structure. In most cases, the impressed current type system is designed to deliver relatively large currents from a limited number of anodes, and the galvanic anode type system is designed to deliver relatively small currents from a large number of anodes.

Each method of applying cathodic protection has characteristics which make it more applicable to a particular problem than the other.

Electric power generating facility intake structures consist of a variety of materials and components which are installed at the cooling water source. Most intake structures consist of trash racks, traveling water screens, and circulating water pumps. Because of the amount of equipment that must be protected from corrosion at power plants, impressed-voltage cathodic protection systems are popular.

A typical intake structure for a 1,000 megawatt generating facility may consist of as many as 8 to 12 intake bays which are typically 10 to 14 feet in width and can be as deep as 80 feet. The water which flows into the intake bay can flow at a rate which is in excess of 2 fps. The corrosion problems encountered for the materials which comprise an intake structure are primarily associated with the coupling of dissimilar metals exposed to a corrosive environment.

Traveling water screen materials can consist of mild steel frames with stainless steel or copper baskets. In addition, various appurtenances which comprise the traveling screen, such as rollers and support frames, can be of different material combinations. Often, the intake structure appurtenances are directly coupled to the reinforcing steel within the intake bay through a common grounding system. This condition also serves to accelerate the potential for corrosion of the traveling water screens, trash racks, and circulating water pumps.

Corrosion control considerations for intake structures are usually limited to the application of cathodic protection for corrosion control. Elaborate protective coating systems are usually not cost effective as a means of corrosion control due to the unique shapes and the physical movement of the traveling screens. The flow of aggressive water against the surfaces of the trash racks, traveling screens, and circulating water pumps will have a detrimental effect on the adhesion of the coating and can result in rapid deterioration of any protective coating system.

The application of an effective cathodic protection system is the only viable method of mitigating corrosion of intake structure components. As previously mentioned, protective coatings can only be somewhat successful in controlling corrosion and frequent repairs to the coating system would be required to render an effective approach. These factors can only contribute to an expensive maintenance program. As in the case of condenser water boxes, the impressed current cathodic protection system is usually the preferred approach for effective corrosion control for traveling screens and trash racks. Galvanic anode systems would require a large volume of anode material to provide a sufficient amount of current to achieve protection and maintain a reasonable cathodic protection system life. Often, the amount of material required would be prohibitive in terms of cost and would require intricate mounting procedures to install the anodes in close proximity to the structures for which protection is required. In most cases, sacrificial anode material mounted on traveling water screen frames would result in too much weight for the screen motor drive assembly and would severely impede the operating characteristics of the traveling water screen.

Typical current requirements can range from 30 to 100 amperes per bay depending upon the width of the traveling water screen and the depth of the submerged area. Typical cathodic protection materials utilized for the protection of intake structure components are high silicon cast iron, 2% lead silver alloy, platinum niobium, and platinum titanium with recent consideration being given to the feasibility of utilizing LIDA® anodes for this application.

Because of the nature of the intake structure components and because of the complexity of the structure and the intricacy of the materials involved, a close coupling of the anode material to the cathode is usually necessary in order to achieve the proper level of protection. One of the most common schemes for achieving corrosion control has been the utilization of perforated pipes which are stacked in the intake bay to house the anode materials which are suspended within the pipes. This approach allows for the physical mounting of the anodes in close proximity to the structure as well as for the proper current distribution. The anodes are protected against physical damage and, with the pipes extending from the intake floor down to the bay floor, the anodes can be easily retrieved for routine inspection as well as for replacement upon useful consumption of the anode material.

APPENDIX

ABBREVIATIONS AND ACRONYMS

ABS, Acrylonitrile-butadiene-styrene

AC, Alternating Current

ACI, Alloy Casting Institute

AFS, American Foundryman's Society

AFBMA, Anti-Friciton Bearing Manufacturers Association

AGMA, American Gear Manufacturers Association

AIChE, American Institute of Chemical Engineers

AISC, American Institute of Steel Construction

AISI, American Iron and Steel Institute

ANSI, American National Standard Institute

API, American Petroleum Institute

APWA, American Public Works Association

AR, Abrasion Resistant

ASA, American Standards Association

ASCE, American Society of Civil Engineers

ASME, American Society of Mechanical Engineers

ASTM, American Society for Testing and Materials

AWG, American Wire Gage

AWS, American Welding Society

AWWA, American Water Works Association

BAT, Best Available Technology

BBL, Barrel

BG, Birmingham gage

BHN, Brinell Hardness Number

BHP, Brake Horsepower

BIPM, International Bureau of Weights and Measures

BLS, Bureau of Labor Statistics

BOD, Biochemical Oxygen Demand

BSI, British Standards Institute

B&S, Brown & Sharpe

BTU, British Thermal Unit

BWG, Birmingham Wire Gage

BWR, Boiling Water Reactor

CAD, Computer Aided Design

CAM, Computer Aided Manufacture

CDS, Cold Drawn Steel

CEMA, Conveyor Equipment Manufacturers Association

CDS, Cold Drawn Steel

C&F, Cost and Freight

CFR, Code of Federal Regulations

CFS, Cubic Feet per Second

CI, Cast Iron

CIF, Cost, Insurance and Freight

CPM, Critical Path Method

CPVC, Chlorinated Polyvinyl Chloride

CRS, Cold Rolled Steel

CSA, Canadian Standards Association

CSO, Combined Sewer Overflow

DC, Direct Current

DFT, Dry Film Thickness

DI, Deionization

DI, Ductile Iron

DIN, Deutsche Industrie Normal

DO, Dissolved Oxygen

DOE, Department of Energy

DP, Differential Pressure

DP, Dimetral Pitch

DWF, Dry Weather Flow

EEI, Edison Electric Institute

EPA, Environmental Protection Agency

EPRI, Electric Power Research Institute

ERDA, Energy Research and Development Administration

FAS, Free Alongside Ship

FDA, Food and Drug Administration

FHCS, Flat Head Cap Screw

FmHA, Farmers Home Administration

FOB, Free On Board

FPM, Feet per minute

FPS, Feet Per Second

FRP, Fiberglass Reinforced Plastic

FWPCA, Federal Water Pollution Control Act

GPM, Gallons Per Minute

G&A, General & Administration

HDPE, High Density Polyethylene

HHCS, Hex Head Cap Screws

HOA, Hand-Off-Automatic

HRS, Hot Rolled Steel

HT, Heat Treat

HWL, High Water Level

I/A, Innovative and Alternative

ICC, Interstate Commerce Commission

ICC, International Chamber of Commerce

ICE, Institute of Civil Engineers

ID, Inside Diameter

IEC, International Electrotechnical Commission

IEEE, Institute of Electrical & Electronic Engineers

IES, Illuminating Engineers Society of North America

IFI, International Fasteners Institute

I/O, Input/Output

ISO, International Standards Organization

IX, Ion Exchange

JIC, Joint Industrial Council

LC, Letter of Credit

LPG, Liquified Petroleum Gas

LWL, Low Water Level

MGD, Million Gallons per Day

MG/L, Milligrams per Litre

MIL, Military Specification

MKS, Meter-Kilogram-Second

MPN, Most Probable Number

MSF, Multi-Stage Flash Evaporation

MUD, Municipal Utility District

MUS, Minimum Ultimate Strength

MW, Megawatt

NACE, National Association of Corrosion Engineers

NACM, National Association of Chain Manufacutrers

NBS, National Bureau of Standards

NEC, National Electrical Code

NEMA, National Electrical Manufacturers Association

NFPA, National Fire Protection Association

NPDES, National Pollution Discharge Elimination System

NPSH, Net Positive Suction Head

NPT, National Pipe Thread

NRDC, National Resources Defense Act

NSF, National Science Foundation

NTIS, National Technical Information Service

OD, Outside Diameter

OEM, Original Equipment Manufacturer

OHL, Overhung Load

O&M, Operation and Maintenance

ORNL, Oak Ridge National Laboratory

OSHA, Occupational Safety and Health Act

OSW, Office of Saline Water

OTS, Office of Technical Services

PC, Personal Computer

PD, Pitch Diameter

PE, Professional Engineer

PERT, Program Evaluation Review Technique

P&ID, Process and Instumentation Diagram

PIV, Positive Infinitely Variable

PH, Precipitation Hardened

PHD, Pead Hourly Demand

PLC, Programmable Logic Controller

POTW, Publically Owned Treatment Works

PPM, Parts Per Million

PSI, Pounds Per Square Inch

PSIA, Pounds Per Square Inch, absolute

PSIG, Pounds Per Square Inch, gage

PUC, Public Utilities Commission

PVC, Polyvinyl Chloride

PWR, Presurrized Water Reactor

Q, Flow

QA, Quality Assurance

QC, Quality Control

RACT, Reasonably Available Control Technology

RAS, Return Activated Sludge

RBC, Rotating Biological Contactor

RCRA, Resource Conservation and Recovery Act

R&D, Research and Development

RFP, Request for Proposal

RFQ, Request for Quotation

RMA, Rubber Manufacturers Association

RPM, Revolutions Per Minute

RO, Reverse Osmosis

SAE, Society of Automotive Engineers

SCR, Silicon Controlled Rectifier

SDWA, Safe Drinking Water Act

SF, Service Factor

SHCS, Socket Head Cap Screw

SI, Systeme International

SIC, Standard Industrial Classification

SME, Society of Manufacturing Engineers

SS, Suspended Solids

SSPC, Steel Structures Painting Council

SST, Stainless Steel

SVI, Sludge Volume Index

SWG, Standard (British) Wire Gage

TBE, Thread Both Ends

TDH, Total Dynamic Head

TEFC, Totally Enclosed Fan Cooled

TOE, Thread One End

TSS, Total Suspended Solids

TWL, Top Water Level

TWS, Traveling Water Screen

UF, Ultrafiltration

UHMW, Ultrahigh-molecular-weight

UL, Underwriters Laboratories

UNS, Unified Numbering System

USDA, US Department of Agriculture

USGS, US Geological Survey

USPHS, US Public Health Service

USSG, United States Standard Gage

UV, Ultraviolet

V, Volt, Velocity, or Volume

VFD, Variable Frequency Drive

VHN, Vickers Hardness Number

WHO, World Health Organization

W&M, Washburn & Moen

WPCF, Water Pollution Control Federation

WPCP, Water Pollution Control Plant

WQA, Water Quality Act of 1987

WS, Welded Steel

WTP, Water Treatment Plant

WWEMA, Water and Wastewater Equipment Manufacturers Association

WWTP, Wastewater Treatment Plant

XP, Explosion Proof

CONVERSION TABLES

TO CONVERT:	*MULTIPLY BY:*	*TO OBTAIN:*
acre-feet	43,560	cubic feet
acre-feet	325,900	gallons
bars	14.5	pounds/square inch
centimeters	0.003281	feet
centimeters	0.3937	inches
centimeters	10	millimeters
centimeters	10,000	microns
centimeters/second	1.969	feet/minute
centimeters/second	0.03281	feet/second
centimeters/second	0.6	meters/minute
centimeters/second	0.02237	miles/hour
centimeters/second	0.036	kilometers/hour
cubic centimeters	0.06102	cubic inches
cubic centimeters	1,000,000	cubic meters
cubic centimeters	26,420	gallons
cubic feet	1,728	cubic inches
cubic feet	0.02832	cubic meters
cubic feet	0.03704	cubic yards
cubic feet	7.48052	gallons (U.S, liquid)
cubic feet	28.32	liters
cubic feet/minute	0.1247	gallons/second
cubic feet/minute	0.4702	liters/second

TO CONVERT:	*MULTIPLY BY:*	*TO OBTAIN:*
cubic feet/minute	62.43	pounds water/minute
cubic feet/second	0.646317	million gallons/day
cubic feet/second	448.831	gallons/minute
cubic meters	35.31	cubic feet
cubic meters	61,023	cubic inches
cubic meters	1.308	cubic yards
cubic meters	264.2	gallons(U.S.liquid)
cubic yards	27	cubic feet
cubic yards	0.7646	cubic meters
cubic yards	202	gallons(U.S. liquid)
cubic yards/minute	0.45	cubic feet/second
cubic yards/minute	3.367	gallons/second
days	86440	seconds
days	1,440	minutes
degrees (angle)	0.01745	radians
degrees/second	0.01745	radians/second
degrees/second	0.1667	revolutions/minute
feet	30.48	centimeters
feet	0.3048	meters
feet	304.8	millimeters
feet	12,000	mils
feet of water	0.03048	kilograms/sq. centimeter
feet of water	304.8	kilograms/sq. meter
feet of water	62.43	pounds/sq. foot
feet of water	0.4335	pounds/sq. inch
feet/minute	0.5080	centimeters/second
feet/minute	0.01667	feet/second
feet/minute	0.3048	meters/minute
feet/second	30.48	centimeters/second
feet/second	18.29	meters/minute
foot-pounds	0.1383	kilogram-meters
foot-pounds/minute	0.01667	foot-pounds/second
foot-pounds/minute	0.0000303	horsepower
foot-pounds/minute	0.0000226	kilowatts
foot-pounds/second	4.6263	btu/hour
foot-pounds/second	0.07717	btu/minute
foot-pounds/second	0.001818	horsepower
foot-pounds/second	0.001356	kilowatts
gallons	3,785	cubic centimeters
gallons	0.1337	cubic feet
gallons	231	cubic inches

TO CONVERT:	*MULTIPLY BY:*	*TO OBTAIN:*
gallons	0.003785	cubic meters
gallons	0.004951	cubic yards
gallons	3.785	liters
gallons (U.S.)	0.83267	gallons (imperial)
gallons of water	8.337	pounds of water
gallons/minute	0.002228	cubic feet/second
gallons/minute	0.06308	liters/second
gallons/minute	8.0208	cubic feet/hour
grams	0.001	kilogram
grams	1,000	milligrams
grams	0.00205	pounds
grams/centimeter	0.000056	pounds/inch
grams/cu centimeter	62.43	pounds/cubic foot
grams/cu centimeter	0.03613	pounds/cubic inch
grams/liter	8,345	pounds/1,000 gallons
grams/liter	0.062427	pounds/cubic foot
grams/sq centimeter	2.0481	pounds/square foot
horsepower	42.44	btu/minute
horsepower	33,000	foot-pounds/minute
horsepower	550	foot-pounds/second
horsepower	0.7457	kilowatts
horsepower	745.7	watts
horsepower-hours	1,980,000	foot-pounds
horsepower-hours	0.7457	kilowatt-hours
inches	2.54	centimeters
inches	0.0254	meters
inches	25400	microns
inches	25.4	millimeters
inches	1,000	mils
kilograms	1,000	grams
kilograms	2.2046	pounds
kilograms	0.0009842	tons (long)
kilograms	0.001102	tons (short)
kilograms/cu. meter	0.06243	pounds/cubic foot
kilograms/cu. meter	0.00003613	pounds/cubic inch
kilograms/meter	0.672	pounds/foot
kilograms/sq. cm.	2,048	pounds/square foot
kilograms/sq. cm.	14.22	pounds/square inch
kilograms/sq. meter	0.2048	pounds/square foot
kilograms/sq. meter	0.001422	pounds/square inch
kilometers/hour	27.78	centimeters/second

TO CONVERT:	*MULTIPLY BY:*	*TO OBTAIN:*
kilometers/hour	54.68	feet/minute
kilometers/hour	0.9113	feet/second
kilometers/hour	16.67	meters/minute
kilometers/hour	0.6214	miles/hour
kilopascals	0.145018	pounds/square inch
kilowatts	1.341	horsepower
kilowatts	73.76	foot-pounds/second
kilowatt-hours	2,655,000	foot-pounds
kilowatt-hours	1.341	horsepower-hours
liters	1,000	cubic centimeters
liters	0.03531	cubic feet
liters	61.02	cubic inches
liters	0.001	cubic meters
liters	0.2642	gallons (U.S. liquid)
liters/minute	0.0005886	cubic feet/second
liters/minute	0.004403	gallons/minute
$\log_{10} n$	2.303	ln n
ln n	0.4343	$\log_{10} n$
meters	100	centimeters
meters	3.281	feet
meters	39.37	inches
meters	1,000	milimeters
meters/minute	1.667	centimeters/second
meters/minute	3.281	feet/minute
meters/minute	0.05468	feet/second
meters/minute	0.03728	miles/hour
meters/second	196.8	feet/minute
meters/second	3.281	feet/second
meters/second	3.6	kilometers/hour
meters/second	2.237	miles/hour
meter-kilogram	7.233	pound-feet
microns	1,000,000	meters
microns	0.0000394	inches
miles	1.609	kilometers
miles	5280	feet
miles/hour	44.7	centimeters/second
miles/hour	88	feet/minute
miles/hour	1.467	feet/second
miles/hour	1.6093	kilometer/hour
miles/hour	26.82	meters/minute
miles/hour	0.01667	miles/minute

TO CONVERT:	*MULTIPLY BY:*	*TO OBTAIN:*
milligrams	0.001	grams
milligrams/liter	1.0	parts/million
milliliters	0.001	liters
millimeters	0.1	centimeters
millimeters	0.003281	feet
millimeters	0.03937	inches
millimeters	0.001	meters
millimeters	39.37	mils
million gallons/day	1.54723	cubic feet/second
mils	0.00254	centimeters
mils	0.001	inches
parts/million	1	milligrams/liter
parts/million	8.345	pounds/million gallons
pounds of water	0.01602	cubic feet
pounds of water	27.68	cubic inches
pounds of water	0.1198	gallons
pound-feet	0.1383	meter-kilograms
pounds/cubic foot	16.02	kilograms/cubic meter
pounds/cubic feet	0.0005787	pounds/cubic inch
pounds/cubic inch	1728	pounds/cubic foot
pounds/feet	1.488	kilograms/meter
pounds/square inch	2.307	feet of water
pounds/square inch	703.1	kilograms/square meter
pounds/square inch	144	pounds/square foot
pounds/square inch	0.0703	kilograms/sq centimeter
pounds/square inch	6.8957	kilopascals
pounds/square foot	0.01602	feet of water
pounds/square foot	4.882	kilograms/square meter
pounds/square foot	0.006944	pounds/square inch
radians	57.296	degrees
radians/second	9.549	revolutions/minute
revolutions/minute	6	degrees/second
square centimeters	0.001076	square feet
square centimeters	0.1550	square inches
square centimeters	0.0001	square meters
square feet	0.00002296	acres
square feet	929	square centimeters
square feet	144	square inches
square feet	0.0929	square meters
square feet	0.1111	square yards
square inches	6.452	square centimeters

TO CONVERT:	*MULTIPLY BY:*	*TO OBTAIN:*
square inches	0.006944	square feet
square inches	645.2	square millimeters
square meters	0.0002471	acres
square meters	10000	square centimeters
square meters	10.76	square feet
square meters	1550	square inches
square meters	1.196	square yards
square millimeters	0.00155	square inches
square yards	9	square feet
square yards	0.8361	square meters
tons (long)	1016	kilograms
tons (long)	2240	pounds
tons (long)	1.12	tons (short)
tons (metric)	1000	kilograms
tons (metric)	2205	pounds
tons (short)	907.18	kilograms
tons (short)	2000	pounds
tons (short)	0.89287	tons (long)
tons (short)	0.9078	tons (metric)
tons of water/24 hrs	83.33	pounds of water/hour
tons of water/24 hrs	0.16643	gallons/minute
tons of water/24 hrs	1.3349	cubic feet/hour

OPEN AREA EQUATIONS

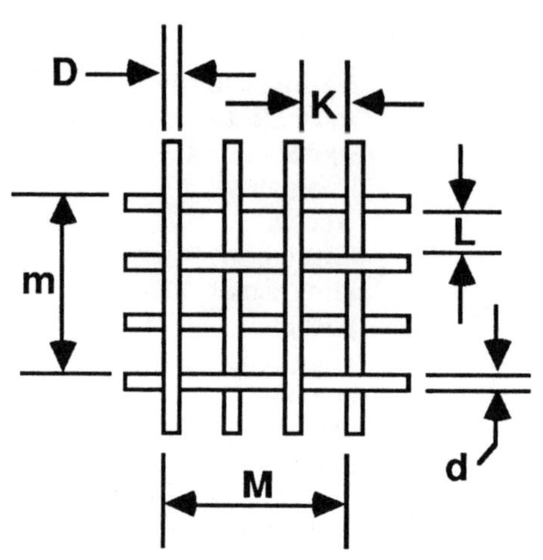

Where:

POA = percent open area

d = shute wire diameter

D = warp wire diameter

K = width of opening (inches)

L = length of opening (inches)

m = shute wires per lineal inch

M = warp wires per lineal inch

For Plain, Twilled or Rectangular:

$$POA = ((K \times L) / (K + D) (L + d)) \times 100$$

or,

$$POA = (1 - MD) (1 - md) \times 100$$

To convert meshes per inch to space openings:

$$K = (1 / M) - D$$

or,

$$L = (1 / m) - d$$

To convert space openings to meshes per inch:

$$M = 1 / (K + D)$$

or,

$$m = 1 / (L + d)$$

RIGHT
or
LEFT HAND?

"Right-hand" and "left-hand" terminology is commonly used when discussing screening equipment. These terms are used to describe an entire screen unit as well as individual screen components. It is important that a universally understood point of reference be used when describing a screen.

Screen Unit The "right-hand" side of the screen is that side of the screen that is on the right when the screen is viewed from the front, with the viewers back to the flow.

Screen Chain Not all chains or attachments are reversible in the direction of travel, in operation over sprockets or in tracks. Therefore, many attachments must be made in both right and left-hand configurations so they may be applied to conveyors requiring two parallel strands of chain.

Right-hand or left-hand identification is made on the basis of an individual link by holding the attachment link so that the surface which contacts the sprocket or track is down, and the closed end of an offset-sidebar chain (or the trailing end of a straight-sidebar link) is away from your body. Under these conditions, if the attachment extends to the right, it is a right-hand attachment; if to the left, it is a left-hand attachment.

SCREENING EQUIPMENT MANUFACTURERS

The following list of manufacturers includes the company name, address, phone number and the type of screens that are known to be currently offered. It is recommended that the manufacturers be contacted to confirm their offerings.

B = Bar Screens
C = Comminutors
D = Drum/Disc Screens
F = Fine Screens
M = Microscreens
R = Trash Rakes
T = Traveling Water Screens

Acme Engineering Company **T**
397 East 9th Avenue
Cordele, Georgia 31015
Phone (912) 273-5590

Amwell, Inc. **B**
1740 Molitor Road
Aurora, Illinois 60505-9990
Phone (312) 898-6900

Ashbrook-Simon-Hartley **B**
P.O. Box 16327
Houston, Texas 77222
Phone (713) 449-0322

Atlas Polar, Hercules Division **R**
P.O. Box 160, Postal Station "O"
Toronto, Canada M4A 2N3
Phone (416) 751-7740

Beaudrey Corporation **B, D, R, T**
333 Sylvan Avenue
Englewood Cliffs, New Jersey 07632
Phone (201) 871-1660

E. Beaudrey & Company **B, D, R, T**
14, bd Ornano 75018
Paris, France
Phone 257.14. 35

Chicago Pump Products, **C**
Yeomans Chicago Corp.
1999 North Ruby Street
Melrose Park, Illinois 60160
Phone (312) 344-4960

Cook Screen Division **F**
LSE Technologies
4848 Franklin Avenue
Cincinnati, OH 45212
Phone (513) 531-6555

Cross Machine, Inc. **R**
P.O. Box 529
Berlin, New Hampshire 03570
Phone (603) 752-6111

Davy Bamag GmbH **B, D, R, T**
Postfach 460
D-6308 Butzbach
Germany
Phone 06033/831

Disposable Waste Systems **C**
2600 South Gurnsey Street
Santa Ana, CA 92707
Phone (714) 751-8100

Duperon Corporation **R**
5693 Becker Road
Saginaw, Michigan 48601
Phone (517) 754-8800

E&I Corporation **B, T**
5655 N. High Street, Suite 111
Worthington, OH 43085
Phone (614) 846-8580

Eimco Process Equipment, Inc. **B, F**
P.O. Box 300
Salt Lake City, Utah 84110-0300
Phone (801) 526-2000

Enviro-Care Company **B, C**
5614 W. Grand Avenue
Chicago, Illinois 60639
Phone (312) 745-7773

Enviroquip, Inc. **B**
P.O. Box 9069
Austin, Texas 78766
Phone (512) 836-1614

Envirex, Inc. **B, D, M, R, T**
1901 South Prairie Avenue
Waukesha, Wisconsin 53186
(414) 547-0141

Electromecanique Verbandt Andre **B**
Roodborstjeslaan 18
8300 Knokke-Heist
Belgium
Phone 050/60 73 43

Fairfield Service Company **B, R, T**
240 North Boone Avenue
Marion, Ohio 43302
Phone (614) 387-3335

Farm Pump & Irrigation **T**
P.O. Box 845
Shafter, CA 92649
Phone (805) 589-6901

FMC Corporation **B, D, F, R, T**
Material Handling Systems Division
400 Highpoint
Chalfont, Pennsylvania 18914
Phone (215) 822-4300

Franklin Miller **B, C**
60 Okner Parkway
Livingston, New Jersey 07039
Phone (201) 535-9200

Hans Huber GmbH **B, F**
MariahilfstraBe 3-5
D-8434 Berching
West Germany
(08462) 201-0

Hawker Siddeley Brackett Ltd. **B, D, F, M, R, T**
Severalls Lane, Colchester
Essex CO4 4PD England
Phone (0206) 852121

Hellmut Geiger GmbH & Co. **B**
Hardeckstrabe 3
7500 Karlsruhe 21
West Germany
Phone 07 21/5987-0

Hendrick Manufacturing Company **F**
7th Ave. & Clidco Drive
Carbondale, Pennsylvania 18407
Phone (717) 282-1010

Hewitt Machine Company **D**
P.O. Box 6888
Neenah, WI 54957
(414) 722-7713

Hubert Water Systems BV **B, D, M, R, T**
Kooyweg 21
8715 EP Stavoren
The Netherlands
Phone (31) 5149-1625

Hycor Corporation **B, D, F**
29850 North Highway 41
Lake Bluff, Illinois 60044
Phone (312) 473-3700

Infilco Degremont, Inc. **B, R**
P.O. Box 29599
Richmond, Virginia 23229
Phone (804) 281-7600

Jeffrey Division, **B, R, T**
Dresser Industries, Inc.
P.O. Box 387
Woodruff, South Carolina 29388
Phone (803) 476-7523

Johnson Screens **F**
P.O. Box 64118
St. Paul, Minnesota 55164
Phone (612) 636-3900

Lakeside Equipment Corporation **B, F, M**
1022 E. Devon Avenue
Bartlett, Illinois 60103
Phone (312) 837-5640

Longwood Engineering Company Ltd. **B, R**
Parkwood Mills, Longwood
Huddersfield, West Yorkshire
HD3 4TP England
Phone (0484) 642011

Mather & Platt Limited **M**
Park Works
Manchester UK M10 6BA
England
Phone (061) 205 2321

McGinnes/Royce, Inc. **B, D, R, T**
P.O. Box 34543
Houston, Texas 77234
Phone (713) 485-4371

John Meunier, Inc. **B, F, R**
6290 Perinault
Montreal, Quebec
Canada H4K 1K5
Phone (514) 334-7230

Ossberger Turbines, Inc. **B, R**
5709 South Laburnum Avenue
Richmond, Virginia 23231
Phone (804) 226-9180

Ossberger Turbinenfabrik GmbH **B, R**
Postfach 425
D-8832 Weissenburg/Bavaria
West Germany
Phone 0 91 41-40 91

Parkson Corporation **B, F**
P.O. Box 408399
Fort Lauderdale, Florida 33340-8399
Phone (305) 947-6610

Passavant-Werke AG & Company **B, R, T**
D-6209 Aarbergen 7
West Germany
Phone 06120/281

Perry Engineering **B**
Railway Terrace,
Mile End South
Adelaide, South Australia
Phone (08) 3521777

Pro-Ent, Inc. **T**
P.O. Box 23611
Jacksonville, Florida 32241
Phone (904) 737-3536

Purator Klaeranlagen **B**
Goldeggasse 2
A-1040 Vienna
Austria
Phone 222-6567850

Ranney Company **F**
P.O. Box 729
Westerville, OH 43081
Phone (614) 882-3104

Schloss Engineered Equipment **B**
10555 E. Bartmouth Ave., Suite 230
Aurora, CO 80014
Phone (303) 695-4500

Schreiber Corporation, Inc. **B**
100 Schreiber Drive
Trussville, Alabama 35173
Phone (205) 655-7466

Serck Baker, Inc. **F**
5352 Research Drive
Huntington Beach, CA 92649
Phone (714) 898-3474

Smalley Excavators, Inc. **R**
71 Hartford Turnpike South
Wallingford, CT 06492
Phone (203) 265-9352

Smith & Loveless, Inc. **B**
14040 Sante Fe Trail Drive
Lenexa, Kansas 66215
Phone (913) 888-5201

Sprout-Bauer, Inc. **F**
P.O. Box 539
Springfield, Ohio 45501
Phone (513) 390-3400

SWECO, Inc. **F**
8029 US Highway 25
Florence, Kentucky 41042
Phone (606) 371-4360

UBE Industries **B, R, T**
ARK Mori Building, 12-32 Akasaka
Minato-ku, Tokyo, 107 Japan
Phone (03) 505-3311

Vulcan Industries, Inc. **B**
212 South Kirlin Street
Missouri Valley, Iowa 51555
Phone (712) 642-2755

Whitehead & Poole Limited **B**
P.O. Box 9
Radcliffe, Manchester
M26 9NU, England
Phone (061) 723 3821

Wiesemann Engineering Inc. **B, F**
P.O. Box 10037
Largo, Florida 33543
Phone (813-535-4495

William Green (Ecclesfield), Ltd. **B, D, R, T**
Ecclesfield, Sheffield S30 3YS
England
Phone (0742) 467201

Zenith-Abwassertechnik **R**
Postfach 1160
D5908 Neukirchen/Siegerland
West Germany
Phone 02735/74-0

Zimpro/Pasavant **B**
301 W. Military Road
Rothschild, WI 54474
Phone (715) 359-7211

GLOSSARY

Alligatoring A surface defect common to coal-tar paints caused by uneven hardening of paint film, stimulated by the suns rays. Results in the upper layer of paints surface cracking and slipping over softer lower stratum.

Alloy Material with metallic properties composed of two or more elements of which at least one is a metal.

Annealing The heating and controlled cooling of a solid material to reduce hardness, improve machinability, or obtain desired mechanical, physical, or other properties.

Anode The positive electrode where current leaves the solution.

Approach Velocity The average water velocity of fluid in a channel upstream of a screen or other obstruction.

Aqua-guard Trademark for Parkson, Corp. filter screen.

Aquifer A geological formation containing large quantity of water.

Auto-Rake Brand name for Franklin Miller Co. hydraulically driven, reciprocating rake bar screen.

B-10 Life The "rated life" that defines the number of revolutions that 90% of a group of identical bearings will complete or exceed before first evidence of fatigue develops.

Babbitted Bearing A bearing using babbitt, a tin or lead based alloy, as an anti-friction lining.

Backlash Movement (if any) of chain along the pitch line of the sprocket when the direction of chain travel is reversed.

Barry Rake Trade name of Cross Machine, Inc. trash rake.

Bandscreen Another common term for a Traveling Water Screen.

Barminutor Brand name of Yeomans Chicago Corp. combination bar screen and comminuting device.

Basket The individual screening elements used in a Traveling Water Screen consisting of a wire mesh panel and its structural frame. Also called a "tray".

BCL Screen Brand name of Hycor Corp. "Back-Cleaned Leader" screen.

Bioflush Brand name of Beaudrey Corp. trash rake.

Bitumastic Brand name of Koppers Company line of bituminous coatings.

Blakeborough Trade name of William Green, Ltd.

BOD Biochemical oxygen demand. A standard measure of wastewater strength that quantifies the mg/l of oxygen consumed in a stated period of time; usually five days.

Boot The lower, or bottom portion of a screen structure.

Brinell Hardness A hardness rating determined by the surface area of the cross section of an indentation made by a 1 cm. steel ball pressed into a sample material with a standard force.

Bubbler System Common terminology for pneumatic type differential level controller.

Bulkhead A partition of wood, rock, concrete or steel used for protection from water.

Bushings, chain The bearing surface for pin rotation when chain articulates over a sprocket. Bushings also provide the bearing surface for chain rollers or sprocket contact in rollerless chain.

Calender The process of passing wire cloth through a pair of heavy rollers to reduce its thickness, or flatten it to produce a smooth surface.

Cable cylinder A hydraulic or pneumatic cylinder that utilizes a cable and pulley(s) system.

Caisson Structure used in underwater work containing air under sufficient pressure to exclude water.

Carbon Steel Steel whose major properties depend on its carbon content, without substantial amounts of other alloying elements.

Carburizing Surface heat treatment process done primarily to increase wear resistance of sprockets or chain pins, rollers, and bushings.

Catenary A curve formed by a chain hanging freely from two points not in the same vertical line.

Cathode The negative electrode where the current leaves the solution.

Cathodic protection The use of an impressed current to prevent or to reduce the rate of corrosion of a metal in an electrolyte by making the metal the cathode for the impressed current.

Cavitation The formation of gas- or vapor-filled cavities within a liquid, formed by mechanical forces.

Centrifugal Casting A casting process usually used for tubular products where a weighed portion of molten metal is poured into a metal mold while it is being spun on a horizontal-spindle machine.

Chabelco Brand name of Rexnord, Inc. chain products.

Chill A metal or graphite insert embedded in the surface of a sand mold or core, or placed in a mold cavity to increase the cooling rate and produce a wear resistant surface.

Chordal Action The effect produced by the center of a chain joint being forced to follow arcs instead of chords of the sprocket pitch circle.

Clarifier A quiescent tank or basin that is used to remove suspended solids through gravity settling.

Climber Brand name of Infilco Degremont Inc. reciprocating rake bar screen.

Coanda effect The tendency of a liquid coming out of a nozzle or orifice to travel close to the wall contour even if the walls direction of curvature is away from the jets axis.

Cofferdam A temporary dam, usually of sheet piling built to provide access to an area that is normally submerged.

Cogwheel A wheel with teeth around its edge.

Colloid Suspended solids with a diameter less than one micron that can not be removed by sedimentation alone.

Combination Chain Popular chain for heavy duty conveyor and elevator applications. This chain has cast block links with steel pins and connecting bars.

Combined Flow Wastewater flow that consists of storm water and sanitary sewage.

Comminutor A circular screen with cutters that grinds large sewage solids into smaller settleable particles.

Composite A material composed of two materials bonded together with one serving as a matrix surrounding the particles or fibers of the other.

Compound 146 Brand name of FMC Material Handling Systems Division poylurethane sprocket tooth insert.

Cone Screen Brand name of Eimco Process Equipment internally fed rotary fine screen.

Cont-flo Brand name of John Meunier, Inc. reciprocating, back rake bar screen.

Contra-Shear Trademark for Eimco Process Equipment Co. line of screening equipment.

Coping The top, or covering of an exterior masonry wall.

Corten Trade name of US Steel Corp. grade of high strength, low alloy steel with enhanced atmospheric corrosion resistance.

Curtain Wall An external wall that is not load bearing. Usually refers to a wall that extends down below the surface of the water to prevent floating objects from entering a screen forebay. Also called a "skimmer wall".

Delrin Trademark of E.I. Du Pont De Nemours & Co. line of high molecular weight acetal resins.

Detritus, inorganic A heterogeneous mass of fragments of stone.

Detritus, organic Decaying organic matter such as root hairs, stems, and leaves usually found on the bottom of a water body.

Diametral Pitch The ratio of the number of teeth in a gear to the diameter of the pitch circle.

Dimminutor Brand name of Franklin Miller comminutor.

Discostrainer Brand name of Hycor Corp. fine screening device.

Doctor Blade A scraping device used to remove or regulate the amount of material on a belt, roller, or other moving or rotating surface.

Dokwed Brand name of Hubert Water Systems BV back cleaned bar screen.

Domestic Wastewater Wastewater originating from sanitary conveniences in residential dwellings, office buildings and institutions.

Double Crimp A type of wire mesh with corrugations in both the warp and shute wire to lock the wires in position.

Drip Proof Designation for motor enclosure with ventilating openings constructed so that drops of liquids or solids falling on the motor at an angle of 15 degrees, or less will not enter the unit either directly or by running along an inwardly inclined surface.

Durometer hardness The hardness of a material measured with an instrument utilizing a small drill or blunt indenter point under pressure.

Duroy Brand name for McGinnes/Royce Corp. chain material.

Dutch Weave Wire mesh woven similar to "plain weave", except that warp wires are usually larger that shute wires, and shute wires are closely spaced resulting in a dense weave.

Dyna-Grind Brand name of FMC-Material Handling Systems Division screenings grinder.

Effluent Partially or completely treated water or wastewater flowing out of a basin or treatment plant.

Elastic limit The greatest unit stress that a material is capable of withstanding without permanent deformation.

Elastomers Synthetic rubbers of hydrocarbon and polymeric materials similar in structure to plastic resins.

Elbow Rake Brand name of Acme Engineering Corp. hydraulic trash rake.

Electrolyte A substance that dissociates into two or more ions when it dissolves in water.

Embrittlement Reduction in ductility of materials due to exposure to certain environments or temperatures.

Entrainment The incorporation of small organisms, including the eggs and larvae of fish and shellfish, into an intake system.

Estuary The mouth of a river in which the river's current meets the sea's tide.

Expanded Metal An open metal network produced by stamping or perforating sheet metal.

Explosion Proof Designation for a motor or electrical enclosure designed to withstand a gas or vapor explosion within the unit, and to prevent ignition of gas or vapor surrounding the unit by sparks, flashes, or explosions which may occur within the unit.

Feedwater Water which, in the best practice, is demineralized and heated to nearly boiler temperature and deaerated before being pumped into a steam boiler.

Filtester Brand name for Mather & Platt, Ltd. test apparatus used to predict filterability of microscreen fabric.

Fish Ladder A structure that permits fish to bypass a dam. The ladder resembles a staircase alongside the dam spillway. Water flows down the staircase, with each step built high enough to allow the fish to jump successive steps, against the flow.

Flint Rim Trade name of FMC/PTC chill rim cast sprocket.

Flo-Screen Trademark for Enviro-Care Co. screw operated reciprocating bar screen.

Flowminutor Brand name of Enviro-Care Co. comminutor.

Flux The amount of some quantity flowing across a given area per unit time.

Forebay A reservoir at the end of a pipeline or channel.

Forging Plastically deforming metal, usually hot, into desired shapes with compressive force.

Frazil Ice Granular or spike shaped ice crystals that form in supercooled water that is too turbulent to permit coagulation into sheet ice.

Free-Slide Brand name of McGinnes/Royce Corp. wire mesh and tray configuration used on their rear discharge fish screen.

Frontloader Brand name of Schreiber Corp. reciprocating rake bar screen.

Frontrunner Trademark of Dresser/Jeffery Division for reciprocating rake bar screen.

Galling Development of a condition on the live bearing surface of mating parts where excessive friction results in localized welding and a further roughening of contact surfaces.

Galvanic Couple The connection of two dissimilar metals in an electrolyte that results in current flow through the circuit.

Galvanizing An electrolytic or hot dipping process to coat steel products with a coating of zinc.

Geiger Screen Reciprocating rake bar screen named after its manufacturer, also marketed as a "Climber" screen.

Grabber Trademark of Hycor, Corp. reciprocating rake bar screen.

Grit The dense, mineral, suspended matter present in a water or wastewater, such as sand, silt, or cinders.

Grit Chamber A settling chamber used to remove grit from organic solids through sedimentation or an air-induced spiral agitation.

Grout Fluid, or Semi-fluid cement slurry for pouring into joints of brickwork or masonry.

Hazardous Area, Class 1 Locations where flammable gases or vapors may be present in the air in sufficient quantities to produce explosive or ignitable mixtures.

Headloss The difference in water level between the upstream and downstream sides of a screen.

Helical Gear Gear wheels running on parallel axes with teeth cut oblique to the gear axis.

Hevi-Duty Trademark of McGinnes/Royce Corp. screen components.

Holiday Any discontinuity or bare spot in a coated surface.

Hunting Tooth Sprocket A sprocket with two sets of effective teeth arranged so that each set makes contact with the chain with alternate revolutions of the wheel.

Hydrasieve Trademark for Sprout-Bauer, Inc. static screen.

Hydroflush Brand name of Beaudrey Corp. cable operated bar screen.

Hydrorake Brand name of Atlas Polar Co., Hercules Division trash rake.

Hydroscreen Brand name for Hycor, Corp. static screen.

Hypress Brand name of Hycor, Corp. screenings press.

Impingement The entrapment of fish and other marine life against screening media that results when organisms cannot escape the area in front of the screen because of the velocity of the intake stream.

Influent Water or wastewater flowing into a basin or treatment plant.

Invert The lowest point of the internal surface of a drain, sewer or channel at any cross section.

Journal That part of a shaft which is supported by, and turns in a bearing.

Journal Bearing A cylindrical bearing which supports a cylindrical rotating shaft.

Key A precision made metal bar used to transfer torque between sprockets or pulleys, and the shafts on which they are mounted. Keys may be of straight or tapered design, and furnished with a "gib-head" to facilitate removal.

L-10 Life See "B-10 Life".

Laminar Flow A flow situation in which fluid moves in parallel layers, usually with a Reynolds number less than 2000.

Ledward & Beckett Trade name of William Green, Ltd.

Lift Screen Brand name of Envirex, Inc. reciprocating rake bar screen.

Link Belt Trade name of FMC Corporation.

Macrofouling Clogging of a condenser tubesheet with debris and biogrowth.

Meehanite The name of several cast irons, cast for a specific purpose such as strength or corrosion resistance, having different combinations of mechanical and engineering properties.

Mesh (as a number) The number of openings per lineal inch, measured from the center of one wire or bar to a point 1" distant.

Micro-Matic Brand name for Permutit Corp. microscreen.

Micro-Sieve Brand name for Passavant Corp. microscreen.

Milliscreen Brand name of Eimco/Contra-shear internally fed rotary fine screen.

Mobius Brand name of Pro-Ent, Inc. belt screen.

Mohs Hardness A measure of hardness of a material as determined by the width of a scratch depth made under a prescribed load.

Moment of Inertia The sum of products formed by multiplying the mass of each element of a figure by the square of its distance from a specified line.

Monomer The basic molecule of a synthetic resin, or plastic.

Muffin Monster Brand name of Disposable Waste Systems, Inc. line of sewage grinders.

NoCling Trademark of Beaudrey, Corp. screening media tray insert.

No-well Brand name of FMC-Material Handling Systems Division platform mounted traveling water screen.

Nylon Plastic compound that offers excellent load bearing capability, low frictional properties, and good chemical resistance.

Para Cone Brand name of Eimco Process Equipment internally fed rotary fine screen.

Parshall Flume A venturi-type flume used to measure flow.

Passive Screen Intake screening device that does not employ mechanical cleaning.

Passivation The changing of a chemically active surface of a metal to a much less reactive state. Usually done to stainless steel by immersion in an acid bath.

Penstocks A pipe which transports water to a turbine for the production of hydroelectric energy.

Pickling Preferential removal of oxide or mill scale from the surface of a metal by immersion in an acid or an alkaline solution.

Pins, chain Chain pins connect chain links. They are locked in the sidebars so all relative rotation occurs between the pin and the bushing.

Pinion The smaller of a pair of gear wheels, or the smallest wheel of a gear train.

Pintle Chain Chain extensively used for elevating and conveying, consisting of one-piece links cast with two offset sidebars and coupled with steel pins. Lugs prevent turning of the pins in the sidebars, insuring that articulation will occur between the pin and the cored cylinder.

Pitch The length of one link of chain measured from pin centerline to pin centerline.

Pitch Diameter The diameter of the pitch circle of a sprocket or gear.

Plain Weave Wire mesh with warp and shute wires passing over and under the next adjacent wire in both directions.

Planetary Gear Train An assembly of meshed gears consisting of a central gear, a ring gear, and one or more intermediate pinions supported on a revolving carrier.

Plastic Deformation Permanent change in shape or size of a solid body without fracture resulting from resulting from the application of sustained stress beyond the elastic limit.

Polymer A compound of high molecular weight derived by the recurring addition of similar molecules.

Powermatic Brand name of William Green, Ltd. reciprocating rake bar screen.

Powrclean Brand name of Aerators, Inc. multi-raked bar screen, developed by the former Welles Products Corp.

Posirake Brand name of Passavant, Corp. reciprocating rake bar screen.

Precipitation Hardening Heat treating process used for some alloys that involves a solution heat treatment and rapid quenching followed by a low temperature precipitation or aging treatment.

Pressveyor Brand name of Hycor, Inc. hydraulic conveyor press.

Profiling The action of a bar screen cleaning rake as it travels around the "profile" of an obstruction.

Profile wire Terminology used by Hendrick Mfg. to describe their selection of various shapes of screen bars.

Promal Trade name for FMC/PTC high strength pearlitic malleable iron chain material.

Primary Treatment The initial treatment of wastewater, usually by sedimentation and/or screening, to remove suspended solids.

Putrescible Organic matter in a state of decay or decomposition.

Rake-O-Matic Brand name of former BIF Division of General Signal hydraulically operated, reciprocating rake bar screen.

Ranney Intake Brand name of Hydro Group, Inc. surface water intake system utilizing a passive screen/caisson arrangement.

Raw Edge An unselvaged edge on a piece of woven wire mesh.

Reacher Brand name of Schloss Engineered Equipment, Inc. reciprocating rake bar screen.

Red Rubber Trademark for Rubber Millers, Inc. cast urethane bar screen rake and rake tooth segments.

Retrofit The extensive modification/upgrading of an existing screen, or its replacement with a different type of screen.

Rex Trademark of Rexnord, Inc.

Reynolds Number A non-dimensional number that measures the state of turbulence in a fluid system. It is calculated as the ratio of inertia effects to viscous effect.

Ristropf screen Traveling water screen equipped with a fish bucket collection and sluice spray discharge system named for one of its developers, J.D. Ristropf.

Rockwell Hardness A measure of hardness of a material as determined by the depth of indentation made by a 12 degree conical diamond under a prescribed load.

Roller chain Chain equipped with a roller that revolves over a stationary bushing.

Rollers, chain Chain rollers reduce the coefficient of friction in roller chain by rolling rather than sliding, and minimize sprocket scrubbing as chain enters and leaves a sprocket.

Rotamat Brand name of Lakeside Equipment Co. rotary fine screen.

Rotarc Brand name of a John Meunier, Inc. full rotary bar screen.

Rotoclear Brand name of Walker Process Co. microscreen.

Rotopac Brand name of a John Meunier, Inc. screw compactor.

Rotostrainer Brand name of Hycor, Corp. externally fed rotary screen.

Rotoshear Brand name of Hycor, Corp. internally fed rotary screen.

Rotostep Brand name of Hycor, Corp. in-channel rotating step screen.

Ryertex Brand name of fabric/plastic laminate bushing material manufactured by J. T. Ryerson & Son, Inc.

Sacrificial Anode A sacrificial piece of metal, usually zinc or magnesium, that is electrically connected to a more noble metal in an electrolyte. The anode goes into solution at a dis-proportionate, accelerated rate to protect the more noble metal from corrosion.

Sanitary Wastewater Domestic wastewater without storm and surface runoff, that originates from sanitary conveniences.

Schoop Process Process for coating steel that uses a blast of air to spray a mist of molten metal onto the surface to be protected.

Secondary Treatment The treatment of wastewater through biological oxidation after primary treatment.

Section Modulus The ratio of the moment of inertia of the cross section of a beam undergoing flexure to the greatest distance of an element of the beam from the neutral axis.

Sedimentation The removal of settleable suspended solids by gravity in a clarifier.

Service Factor A multiplier that, when applied to the rated power, indicates the permissible power loading that may be carried under the conditions specified.

Selvage A finished edge on wire mesh to prevent its unravelling.

Shear Pin Sprocket A drive sprocket equipped with a shear pin device used to protect equipment from jamming or overloads. The replaceable shear pins transmit the required toque under normal conditions, and fail when overloaded.

Sherardizing A process for protecting iron from corrosion by means of a corrosion resistant layer of zinc on the iron surface.

Shore (Scleroscope) Hardness A measure of hardness by dropping a diamond tipped "hammer" on a material and the rebound taken as an index of hardness.

Shute The horizontal wire in woven wire mesh, also called the "weft" wire.

Sidebars, chain Chain sidebars are the tensile members of a chain.

Skip Cleaning rake.of a bar screen.

Sluice gate Manual or power operated gate used to isolate a channel from flow.

Slush Oil A protective non-drying oil or grease that adheres to steel surfaces, remains soft for prolonged periods, and can be readily removed.

Smooth-tex Brand name of rectangular woven wire mesh offered by McGinnes/Royce Corp.

Sta-Sieve Brand name of SWECO, Inc. static screen.

Stapling The entanglement of stringy or fibrous debris on a mesh or bar rack.

Stellite Trade name of Cabot Corp. for a family of cobalt-based alloys used as a bushing material.

Stoody Manufacturer of a line of wear resistant alloy parts. Commonly used term to refer to their centrifugally cast chrome-nickel-boron bushing material.

Stop Log A removeable wooden, steel or concrete bulkhead which fits in vertical grooves in a channel to stop water flow.

Straightline Brand name of FMC-Material Handling Systems Division cable operated bar screen.

Suboscreen Brand name of Eimco Process Equipement in-channel rotary fine screen.

Supermal Brand name for Dresser/Jeffery high strength pearlitic malleable iron chain material.

Suspended Solids (SS) Solids captured by filtration through a glass wool mat or 0.45 micron filter membrane.

Taper-lock Sprocket Term for sprockets equipped with a split tapered bushing for rigid mounting on a shaft.

Taskmaster Brand name for Franklin Miller shedder.

Tensile strength The maximum tensile load per square unit of cross section that a material is able to withstand.

Tertiary The use of physical, chemical, or biological means to improve secondary effluent quality.

Thermoplastic A high polymer that flows and melts when heated. Scrap may be recovered by remelting and reusing.

Thermosetting plastic A high polymer that sets into a rigid network. These polymers are not reusable.

Thimble, chain Another term used to refer to a chain bushing.

Thru-Clean Brand name of FMC-Material Handling Systems Division back cleaned, multi-rake bar screen.

Tines The tooth or prong of a bar screen cleaning rake.

Toughness The property of a material which enables it to absorb energy while being stressed above its elastic limit but without being fractured.

Totally Enclosed Fan Cooled (TEFC) Designation for motor enclosure that is not airtight, but constructed so as to prevent free exchange of air between the inside and outside of the motor case. Exterior cooling is provided by a fan integral with the machine, but external to the enclosing parts.

Totally Enclosed Nonventilated (TENV) Designation for motor enclosure that is not airtight, but constructed so as to prevent free exchange of air between the inside and outside of the motor case.

Transducer A device that receives energy from one system and retransmits it, often in another form, to another system.

Tray The individual screening elements used in a Traveling Water Screen consisting of a wire mesh panel and its structural frame. Also called a "basket".

Tritor Brand name of FMC-Materials Handling Systems Division combination bar screen and grit removal device.

Triturator Brand name of Envirex, Inc. screenings grinder.

Turbulent Flow A flow situation in which the fluid moves in a random manner, with a Reynolds number usually greater than 4000.

Turnover Seasonal (spring and fall) change that occurs in a lakes thermal gradients, resulting in the circulation of biological and chemical materials.

Twilled Weave Wire mesh woven with each warp and shute wire passing over two and under two of the next adjacent pair of wires.

Ultimate Strength The stress, calculated on the maximum value of the force and the original area of the cross section which causes fracture of the material.

Vee-wire Trademark of Johnson Division of UOP, Inc. wedge-shaped wire.

Velocity Cap The horizontal cap on an vertical, offshore water intake that results in a horizontal inflow, thus reducing fish entrainment.

Vickers Hardness An accurate hardness rating determined by the surface area of the cross section of an indentation made by a pyramidal shaped diamond pressed into a sample material with a standard force.

Warp The vertical wire in woven wire mesh.

Waterbox The chamber at the inlet end of a condenser tubesheet.

Wedgewater Sieve Brand name for Hendrick Mfg. Co. static screen.

Wedgewire General term to describe trapezoidal or v-shaped wire.

Weft The horizontal wire in woven wire mesh, also called the "shute" wire.

Weir An adjustable baffle over which water flows.

Working Load An allowable recommended tensile load for chains used on conveyors, screens, or other applications of low relative speeds.

Worm A shank having at least one complete thread around the pitch surface.

Worm Gear A gear with teeth cut on an angle to be driven by a worm; used to connect non-parallel, non-intersecting shafts.

Yield That stress in a material at which plastic deformation occurs.

Z-metal Trade name for Rexnord, Inc. high strength, pearlitic, malleable iron chain material.

BIBLIOGRAPHY

Catalogs, Brochures and Product Bulletins

"Acme Stationary Trash Rake", Acme Engineering Co., Cordele, GA, Product Bulletin

"Advanced Water Intake and Screening Equipment", Beaudrey Corp., Englewood Cliffs, NJ, Product Bulletin

"Application Report No. 7", Rex Chainbelt Inc., Milwaukee, WI, Bulletin 315-256 (1967)

"Aqua Guard Screen", Parkson Corp., Fort Lauderdale, FL, Bulletin AG-400 (1982)

"Aqua Guard Screen", Parkson Corp., Fort Lauderdale, FL, Bulletin AG-400 (1986)

"Arc Screen Self Cleaning Curved Bar Screen", Infilco Degremont Inc., Richmond, VA, Bulletin DB 814 (1984)

"Automatic Water Screens", J. Blakeborough & Sons Ltd., West Yorkshire, England, Bulletin 309/B (1976)

"Auto-Rake", Franklin Miller, Livingston, NJ, Product Bulletin

"BCL Screen", Hycor Corp., Lake Bluff, IL, Bulletin 878

"The Barber Surf-Rake", H. Barber & Sons Inc., Bridgeport, CT, Bulletin 7182

"Barminutor", Chicago Pump Products, Melrose Park, IL, Bulletin 7641-C

"The Berry Trash Rake", L.H. Berry Inc., Conway NH, Product Bulletin

"E. Beaudrey & CIE", E. Beaudrey & CIE, Paris, France, Product Bulletin

"Brackett EVA Back-Raked Screens", Hawker Siddeley Brackett Ltd., Essex, England, Bulletin 2005

"Brackett Rotating Drum Screens", Hawker Siddeley Brackett Ltd., Essex, England, Product Booklet

"Brush Raked Curved Fine Screen", William Green & Co., Sheffield, England, Bulletin WG/S6-6/82 (1982)

"Carpenter Stainless Steels", Carpenter Technology Corp., Reading, PA, 2/87/15M Catalog (1987)

"Catenary Bar Screens & Catenary Trash Rakes", Dresser Industries Inc., Woodruff, SC, Bulletin 28186 (1986)

"Central Flow Band Screen", Hawker Siddeley Brackett Ltd., Essex, England, Product Bulletin

"Climber Screen", Infilco Degremont Inc., Richmond, VA, Bulletin DB 803 (1982)

"Cont-Flow Bar Screen", John Meunier Inc., Montreal, Quebec, Canada, Bulletin 10M-5-84 (1984)

"Cont-Flo News Reprint", John Meunier Inc., Montreal, Quebec, Canada, Bulletin 10-1M-10-81 (1981)

"Contra-Shear Rotary Screen Technology", EIMCO Process Equipment Co., Salt Lake City, UT, Product Bulletin (1986)

"CPC Rotary Screen", CPC Engineering Corp., Sturbridge, MA, Bulletin CPC 301 (1977)

"Curved Bar Screens", William Green Ltd., Sheffield, England, Bulletin WG/S5-2/83 (1983)

"Discostrainer", Hycor Corp., Lake Bluff, IL, Bulletin 857

"Double & Single Flow Bandscreens", Willian Green Ltd., Sheffield, England, Product Bulletin

"Dresser Frontrunner", Dresser Industries, Woodruff, SC, Product Bulletin

"Drum Screen, Dual-Flow Screen", McGinnes/Royce, Inc., Houston, TX, Product Bulletin (1982)

"Dual-Flow Traveling Water Screens Model 57 Service Instructions", FMC Corp., Chalfont, PA, Booklet 724102

"Electro-Mechanical Rake", Electromecanique Verbandt Andre, Knokke-Heist, Belgium, Product Bulletin

"E.V.A.", Electromecanique Verbandt Andre, Knokke-Heist, Belgium, Product Bulletin 050/78.92 61

"Equipment for the Treatment of Water, Sewage and Industrial Waste", Jeffrey Manufacturing Co., Columbus, OH, Bulletin 1052-B (1974)

"Fine Mesh Screen Retrofits", Beaudrey Corp., Englewood Cliffs, NJ, Product Bulletin

"Flo-Screen", Enviro-Care Co., Chicago, IL, Product Bulletin

"Flowminutor", Enviro-Care Co., Chicago, IL, Product Bulletin 3K84 (1984)

"FMC Traveling Water Screens", FMC Corp., Chalfont, PA, Product Bulletin

"The FS-304 Self-Cleaning Filter Screen", Wiesemann Engineering Inc., Largo, FL, Product Bulletin

"Griger" Hellmut Geiger GmbH & Co., Karlsruhe, West Germany, Product Brochure 11-26 176/9 (1987)

"Getting solids Out ... Now Easy As A, B, C", Bauer Combustion Engineering Inc., Springfield, OH, Product Bulletin

"Guidelines to Water Screening, Plant Design & Layout", E. Beaudrey & CIE, Paris, France, Product Bulletin

"The Grabber Automatic Bar Screen", Hycor Corp., Lake Bluff, IL, Bulletin 872

"The Hendrick Wedgewater Sieve", Hendrick Manufacturing Co., Carbondale, PA, Bulletin HFS 10815M

"Hercules Hydrorake System", Atlas Polar Company Ltd., Toronto, Ontario, Canada, Product Bulletin

"Hydroscreen", Hycor Corp., Lake Bluff, IL, Product Bulletin

"Hypress", Hycor Corp., Lake Bluff, IL, Product Bulletin

"Hyveyor", Hycor Corp., Lake Bluff, IL, Bulletin 874

"Introducing Hawker Siddeley Barackett", Hawker Siddeley Brackett Ltd., Essex, England, Publication 2000

"Involute Panel Bandscreens", William Green Ltd., Sheffield, England, Bulletin WG/S1-2/83 (1983)

"Johnson Surface Water Intake Screens", UOP Inc., St. Paul, MN, Product Bulletin 201081 (1981)

"Johnson Surface Water Intake Screens", UOP Inc., St. Paul, MN, Product Bulletin (1987)

"The Kason Cross-Flo Sieve", Kason Corp., Newark, NJ, Bulletin CF-79

"Level Measuring Systems", Milltronics, Arlington, TX, Bulletin 12/83-1M (1983)

"Link-Belt Four-Rope Bar Screens", FMC Corp., Chalfont, PA, Service Instructions 12240.80

"Link-Belt Traveling Water Screens", FMC Corp., Chalfont, PA, Bulletin 41006

"Lo Flo Radial Bar Screen", William Green Ltd., Sheffield, England, Product Bulletin

"Mechanically Raked Bar Screens", William Green Ltd., Sheffield, England, Bulletin WG/S4-2/83 (1983)

"Mechanical Rake Screen", Ashbrook-Simon-Hartley, Houston, TX, Product Bulletin

"Micromesh Strainer", Lakeside Equipment Corp., Bartlett IL, Bulletin 230 (1977)

"Microscreens", Envirex Inc., Waukesha, WI, Bulletin 315.31 3M-11/84 (1981)

"Microscreens", Hawker Siddeley Brackett Ltd., Essex, England, Ref. No. 2012

"Microstraining", Mather & Platt Ltd., Manchester, England, Publication AP47A (1987)

"Microstraining", Mather & Platt Ltd., Manchester, England, Publication AP51

"Microstraining", Mather & Platt Ltd., Manchester, England, Publication AP52

"Microstraining", Mather & Platt Ltd., Manchester, England, Publication AP53

"Microstraining and its Applications", Mather & Platt Ltd., Manchester, England, Publication AP1/R3 (1986)

"Microstraining Anti-Pollution Systems", Mather & Platt Ltd., Manchester, England, Publication AP47A

"The Milliscreen Treatment Plant", Contra-Shear Developments Ltd., Auckland, New Zealand, Publication

"Mobius Screening System", Pro-Ent, Inc., Jacksonville, FL, Product Bulletin

"Model 572 Level Measurement System", Ametek Inc., Feasterville, PA, Bulletin DB-572 (1987)

"Muffin Monster", Disposable Waste Systems Inc., Santa Ana, CA, Bulletin F2-MM1180

"No-Well Traveling Water Screen", FMC Corp., Chalfont, PA, Bulletin 814101 (1978)

"No-Well Traveling Water Screens Service Instructions", FMC Corp., Chalfont, PA, Booklet 624101

"The Original Chicago Pump Comminutor", Yeomans Chicago Corporation., Melrose Park, IL, Bulletin 7610-D

"Ossberger Trash Rack Cleaning Systems", Ossberger Turbines Inc., Richmond, VA, Product Bulletin

"Parkwood Inlet Screens", The Longwood Engineering Company Ltd., West Yorkshire, England, Publication LE/3/86 (1986)

"Passavant Cable Operated Bar Screen", Passavant-Werke AG & Co., Aarbergen, FRG, Product Bulletin

"Passavant Cooling Water Screening", Passavant-Werke AG & Co., Aarbergen, FRG, Product Bulletin

"Passavant Micro-Sieve", Passavant Corp., Birmingham, AL, Bulletin 1661 (1975)

"Passavant Posirake Model 1230", Passavant Corp., Birmingham, AL, Bulletin 1231 (1975)

"Passavant Travelling Band Screen", Passavant-Werke AG & Co., Aarbergen, FRG, Product Booklet

"Permutit Micro-Matic Rotating Drum Strainers", Permutit Co., Paramus, NJ, Bulletin 6210-982-5M (1982)

"Posirake Bar Screen", Passavant Corp., Birmingham, AL, Product Booklet (1986)

"Powermatic Screen", William Green Ltd., Sheffield, England, Bulletin WG/S3-2/83 (1983)

"The Power of the Roller Chain!", Vulcan Industries Inc., Missouri Valley, IA, Product Bulletin

"Powrclean Bar Screen", Peabody Welles, Roscoe IL, Bulletin PW 0100-1 (1978)

"Preliminary Treatment", Biwater Sewage Treatment, Manchester, England, Product Bulletin

"Pressveyor", Hycor Corp., Lake Bluff, IL, Product Bulletin

"The Purpose and Application of S&L Mechanically Cleaned Bar Screens", Smith & Loveless Engineering Data, Lenexa, KS (1977)

"Radial Bar Screen", William Green Ltd., Sheffield, England, Bulletin WG/S7-3/83 (1983)

"Ranney Water Collector Systems", Hydro Group, Inc., Westerville, OH, Bulletin 1M-11-83 (1983)

"Replacement Parts and Service", Power Plant Service Supply, Union City, CA, Parts List (1987)

"Rex Heavy-Duty Bar Screens", Rex Chain Belt Co., Milwaukee, WI, Bulletin 315-22 (1961)

"Rex Mechanically Cleaned Bar Screens", Envirex Inc., Waukesha, WI, Bulletin 315-21

"Rex Power Transmission and Conveying Components" Rexnord, Inc., Milwaukee, WI, R85 Catalog (1984)

"Rex Traveling Water Screens", Rex Chain Belt Co., Milwaukee, WI, Catalog 147 (1927)

"Rex Traveling Water Screens", Rex Chain Belt Co., Milwaukee, WI, Catalog 187 (1929)

"Rex Traveling Water Screens", Rex Chain Belt Co., Milwaukee, WI, Bulletin 336 (1938)

"Rex Water Screening Equipment", Envirex, Inc., Waukesha, WI, Bulletin 315-321

"Rotamat Screens", Lakeside Equipment Co., Bartlett, IL Product Bulletin

"Rotarc Brush Screens", John Meunier Inc., Montreal, Quebec, Canada, Bulletin 7-2M-10-81 (1981)

"Rotarc Mini Screens", John Meunier Inc., Montreal, Quebec, Canada, Product Bulletin

"Rotary Screens and Microstrainers" William Green, Ltd., Sheffield, England, Bulletin WG/S2-2/83 (1983)

"The Rotopac Screw Compactor", John Meunier Inc., Montreal, Quebec, Canada, Bulletin U-M1-3m-6-83 (1983)

"Rotoshear", Hycor Corp., Lake Bluff, IL, Bulletin 105-R (1983)

"Rotostep", Hyco Corp., Lake Bluff, IL, Bulletin 865

"Rotostrainer", Hycor Corp., Lake Bluff, IL, Bulletin RS 1101 483 (1977)

"RR Continuous Screw Dewatering Presses", AMETEK Inc., Fresno, CA, Bulletin RR-IND (1) (1986)

"Schreiber Frontloader Screen", Schreiber Corp., Trussville, AL, Product Bulletin

"Screening Equipment", UBE Industries Ltd., New York, NY, Product Bulletin

"Screening Equipment", FMC Corp., Chalfont, PA, Book 2587A, (1966)

"Screening Technology by Contra-Shear", Contra-Shear Developments Ltd., Auckland, New Zealand, TechArt 213 L

"Self-Cleaning Trashrack", The Duperon Corp., Saginaw, MI, Product Bulletin

"Service Instructions for Link-Belt Traveling Water Screens", FMC Corp., Chalfont, PA, Service Instruction 2690 (1973)

"Semi-Rotary Bar Screen", William Green Ltd., Sheffield, England, Bulletin WG/S9-3/83 (1983)

"Single and Double Flow Band Screens, Drum & Cup Screens", Ledward & Beckett Ltd., London, England, Product Bulletin

"Single Tank Level Measuring System", Inventron McGraw-Edison Company, Section 9100

"Smalley Trash Rakes", Smalley Excavators Ltd., Lincs. England, Product Bulletin

"Spray Nozzles", FMC Corp., Chalfont, PA, Folder 32301 (1973)

"Spray Nozzle User's Manual", Spraco Inc., Nashua, NH, Catalog 8507 (1985)

"Stormwater Overflow Screens", John Meunier Inc., Montreal, Quebec, Canada, Product Bulletin

"Straightline Bar Screen", FMC Corp., Chalfont, PA, Folder 2945 (1964)

"The Tallest Cont-Flo Bar Screen", John Meunier Inc., Montreal, Quebec, Canada, Product Bulletin (1982)

"Taskmaster", Franklin Miller Inc., Livingston, NJ, Product Bulletin

"Test Results Using Contra-Shear Screens", Contra-Shear Developments Ltd., Auckland, New Zealand, Publication (1983)

"Tetko Fabrication Services Guide", Tetko Inc., Memphis, TN

"There's gold in them there screens.", Hycor Corp., Lake Bluff, IL, Product Bulletin

"Traveling Screens to Protect Fish in Water Intakes Systems", Envirex, Inc. Waukesha, WI, Bulletin 316-300 (1973)

"Traveling Water Screens", FMC Corp., Chalfont, PA, Catalog 2652

"Traveling Water Screens", FMC Corp., Chalfont, PA, Booklet 51009 (1975)

"Traveling Water Screens", FMC Corp., Chalfont, PA, Bulletin 710101 (1987)

"Traveling Water Screens Fish Protection System", FMC Corp., Chalfont, PA, Bulletin 41006 (1974)

"Traveling Water Screen Model 45A Thru-Flow", FMC Corp., Chalfont, PA, Bulletin 41007 (1974)

"Traveling Screen (Differential Level) Control", Drexelbrook Engineering Company, Bulletin 303-600-A (1987)

"Typical Biological Waste Treatment System", Smith & Loveless Inc., Lenexa, KS, Bulletin S-121

"Wastewater Disintegrating and Screening", Franklin Miller Inc., Livingston, NJ, Product Bulletin

"Waste Water Treatment Equipment", Fairfield Service Co., Marion, OH, Product Bulletin

"Wastewater Disentegrating and Screening", Franklin Miller Inc., Livingston, NJ, Booklet

"Water", Davy Bamag GmbH, Butzbach, FRG, Product Bulletin

"Water Screening Systems Delivery Program", Esmil Hubert BV, The Netherlands

"Water Supply, Treatment", VIZGEP, Budapest, Hungary, Product Brochure

"Water Treatment Equipment", Smith & Loveless Inc., Lenexa, KS, Bulletin 1627MG

Publications, Books and General References

Agranoff, J. , Editor, *Modern Plastics Encyclopedia*, McGraw-Hill Inc., New York, NY (1982)

Bell, M.C., Mussalli, Y.G., Richards, R.T., Taft, E.P., Wagner, C.H., Watts, F.J., *Design of Water Intake Structures for Fish Protection*, American Society of Civil Engineers, New York, NY (1982)

Borowiec, A.N., Chack, J.J., Lopez, A.R., Melbinger, N.R., *The Bar Screens of New York*, WPCF Operations Forum (1987)

Brady, G.S. and Clauser, H.R., *Materials Handbook*, McGraw-Hill Inc., New York, NY (1978)

Cravens, J.B., Kormanik, R.A., Wittman,J.W., *Innovative Application of Microscreening in Lagoon Effluent Polishing*, Presented at The Sixth Mid-America Conference on Environmental Engineering Design (1982)

Fletcher, R.A., *Risk Analysis for Fish Diversion Experiments: Pumped Intake Systems*, Transactions of the American Fisheries Society (1985)

Gauthier, H., Jackson, E., LeGrand, A., Simon, R., *Cooling Water Screens and Pumps in French Power Plants*, EPRI Symposium on Condenser Macrofouling Control Technologies: The State of the Art, Hyannis, MA (1983)

Hazen and Sawyer, Engineers, *Process Design Manual for Suspended Solids Removal*, U.S. Environmental Protection Agency, Washington, D.C. (1975)

Kemmer, F.N. (Technical Editor), *Nalco Water Handbook*, McGraw-Hill Inc., New York, NY (1979)

Majewski, W. and Miller, D.C. (Editors), *Predicting Effects of Power Plant Once-through Cooling on Aquatic Systems*, UNESCO, Paris, France (1979)

McGarry, J.A., Lenhart, C.F., *Screening System Makes Downstream Life Simpler*, WATER/Engineering & Management Magazine, June (1984)

Mussalli, Y.G. and Taft, E.P., *Fish Return Systems*, Proceedings from 25th Annual Hydraulics Division Specialty Conference, Texas A&M University, College Station, TX (1977)

Mussalli, Y.G., Project Manager, Stone & Webster Engineering Corp., and Hillman, R.E., Project Manager, BATTELLE, *Guidelines on Macrofouling Control Technology*, U.S. Environmental Protection Agency, Washington, D.C. (1987)

Nakato, T., Gay, G.E., Kennedy, J.F., *Model Tests of Proposed Intake Structures*, Iowa Institute of Hydraulic Research, University of Iowa (1979)

O'Keefe, W., *Intake Technology Moves Ahead*, Power Magazine, New York, NY (1978)

Parker, S.P. (Editor-in-Chief), *McGraw-Hill Dictionary of Scientific and Technical Terms*, McGraw-Hill Inc., New York, NY (1984)

PEDco Environmental Inc., *Economic Analysis of Section 316(b) of FWPCA on the Steam Electric Utility Industry*, U.S. Environmental Protection Agency, Washington, D.C. (1977)

POWER Magazine, *Low-cost intake structure saves on upkeep*, McGraw-Hill, Inc., (1973)

Ray, S.S., Snipes, R.L. and Tomljanovich, D.A., *A State-of-the-Art Report on Intake Technologies*, TVA Power Research Staff, Chattanooga, TN (1976)

Richards, R.T., *Fish Screening-The State of the Practical Art*, Proceedings from 25th Annual Hydraulics Division Specialty Conference, Texas A&M University College Station, TX (1977)

Scott, J.B., *Dictionary of Civil Engineering*, Halsted Press, New York, NY (1981)

Strow, D., *Continuous Operation of Traveling Water Screens*, Unpublished, Hales Corners, WI (1987)

US EPA Municipal Operations Branch, *Field Manual for Performance Evaluation and Troubleshooting at Municipal Wastewater Treatment Facilities*, Washington, D.C. (1978)

Updegraff, K.F., *Microscreens Applied to Wastewater Treatment Pond Effluent*, Presented at USEPA's Field Evaluation of I/A Technologies (1986)

Wahanik, R.J., *Influence of Ice in the Design of Intakes,* Proceedings from ASCE Specialty Conference on Applied Techniques for Cold Environments, Fairbanks, AL (1978)

WPCF and AWWA Joint Committee, *Wastewater Treatment Plant Design, A Manual of Practice*, Lancaster Press Inc., Lancaster, PA (1977)

INDEX

Bar Screen
- Arc, 146-148
- Back Cleaned, 114-120
- Bar Rack, 155-156
- Cable Operated, 125-128
- Catenary, 120-123
- Chain Driven, 139-140
- Chain Lift, 135-139
- Cogwheel, 128-135
- Comminutors, 149-152
- Deadplate, 156
- Debris Disposal, 164-166
- Debris Volume, 163-164
- Fine, 114
- Front Cleaned, 108-114
- Grit/Bar, 110-112
- Housings, 157-158
- Hydraulic, 141-143
- Manual, 149
- Multi-Raked, 107-108
- Rakes, 156-157
- Reciprocating Rake, 124-125
- Screw Driven, 143-146
- Sizing, 159-163

Cathodic Protection, 206-210

Continuous Operation, 79-81

Controls
- Differential Controller, 194-197
- Float Switch, 194
- Hand-Off-Auto, 193

Drum Screen
- Double-Entry, 66-69
- Single-Entry, 69-71

Fine Screen,
- Continuous Element, 169-171
- Disc-Type, 178-179
- Rotary, 171-177
- Static, 179-181

Fish Screen
- Angled, 90
- Barriers, 90
- Flush Mount, 87
- Front Discharge, 82-84
- Horizontal, 88-89
- Passive, 88
- Platform Mount, 87
- Radial Well, 88
- Rear Discharge, 85-86

Frazil Ice, 64-65
Galvanizing, 208
Grit Screen, 110-112
Intake Location, 81
Materials, 198-205
Microscreening
- Applications, 183-184
- Backwash Spray, 187-190
- Drive & Controls, 190-191
- Drum Support, 185-187
- Mesh, 185
- Sizing, 191-192

Paint, 207
Passive Screen, 73-76
Radial Well Screen, 76-77
Revolving Disc Screen, 71-73
Trash Rake
- Cable Operated, 94-99
- Chain Operated, 105-106
- Hydraulic, 99-105

Traveling Water Screen
- Applications, 2-5
- Backup Beam, 32-34
- Belt Screen, 22-25
- Boot Section, 45
- Carry-over, 9, 15
- Chain, 35-36
- Design & Sizing, 54-64
- Double-Entry, 15-16
- Down Spray, 19-20
- Drive, 52-54
- Dual Flow, 9-17
- Framework, 40-47
- Headloss, 62-64
- Housing, 50
- Inclined, 21
- Mesh, 34-35
- Platform Mounted, 25-27
- Retrofit, 17-19
- Single-Entry, 17
- Spray System, 47-50
- Sprockets, 36-40
- Thru Flow, 5-9
- Trays, 27-32
- Trough, 51
- Velocity, 60-62

CARDIFF UNIVERSITY PRIFYSGOL CAERDYDD